TRANSCENDENT PHYSICS

AF722043

Also by James A. Cusumano

Nonfiction

Freedom from Mid-East Oil
(Coauthors: Jerry B. Brown and Rinaldo S. Brutoco)

Cosmic Consciousness:
A Journey to Well-Being, Happiness, and Success
(1st and 2nd editions)

BALANCE: The Business-Life Connection

Life Is Beautiful: 12 Universal Rules

Dance to the Rhythm of the Cosmos:
Tapping into the Infinite Power of Consciousness

Fiction

The Fallen: As Above, So Below

Twin Terror: Good Seed, Bad Seed

The Dialogue: A Journey to Universal Truth

I Can See Clearly: Rise of a Supernatural Hero
(Book I of the Luc Ponti Chronicles)

Seagull's Revenge: Beyond Fear
(Book II of the Luc Ponti Chronicles)

Cosmic Contact: The Next Earth—a New Beginning
(Book III of the Luc Ponti Chronicles)

JAMES A. CUSUMANO

TRANSCENDENT PHYSICS

TOWARD A *NEW* SCIENCE OF CONSCIOUSNESS

REFLECTIONS FOR CONSIDERATION
BY
NOVICES AND EXPERTS

Waterside Productions

Copyright © 2025 by James A. Cusumano
www.JimTheAlchymist.Com

All rights reserved. This book or any portion thereof may not be reproduced or used in any manner whatsoever without the express written permission of the publisher except for the use of brief quotations in articles and book reviews.

First Printing, 2025

ISBN: 978-1-968401-26-9 (print edition)
ISBN: 978-1-968401-27-6 (eBook edition)

Printed in the United States of America

Waterside Productions
2055 Oxford Avenue
Cardiff-by-the-Sea, CA 92007

For Julia . . .

and

for those intrepid scientists and philosophers willing to endure the social, political, and academic reproach attached to developing the science of Transcendent Physics . . .

Contents

Author's Preface

More than 3,000 years ago, the Wisdom Seekers of the East unveiled profound insights into our universe and its fundamental connections to consciousness. It's striking that only recently have modern cosmologists, astrophysicists, philosophers, and quantum physicists begun to consider these ancient findings with a serious eye. Many scientists remain reluctant to accept them, as they challenge established views of reality and the underlying nature of all existence.

This conundrum became evident in the early 20th century during the advent of quantum mechanics when Max Planck, the father of this new science, asserted in an interview with the *London Observer* that consciousness is fundamental—*intrinsic to the universe and the source of all energy and matter.*[1] Although a number of his esteemed colleagues, including several Nobel laureates, did not initially agree with Planck's analysis, they ultimately concurred with his interpretation.[2] It's often a challenge for even the best and brightest to dismiss the status quo.

As an intrinsic entity, consciousness represents the most fundamental aspect of the universe's architecture. It cannot be divided—unlike normal matter, which can be dissected into smaller components—molecules, atoms, subatomic particles, and quarks, the smallest known constituents of matter.

Why, then, has the intrinsic nature of consciousness and its role as the source of all matter and energy been inadequately explored in subsequent research? Why are scientists only now reluctantly and seriously engaging with these concepts? I believe there are two primary reasons.

First, with stunning foresight, the Wisdom Seekers posited that consciousness is the only cosmic constituent truly intrinsic to the universe. They arrived at this conclusion through practices that could be classified as *Transcendent Physics*—a discipline only broadly accessible to those who have significantly elevated their level of consciousness to a critical part of their research pursuits. The Wisdom Seekers excelled in this practice.

Modern scientists often find Transcendent Physics resistant to the quantitative principles of the *Scientific Method*. Yet, many *qualitative* insights derived from this new science far surpass the import and potential of the *quantitative* outcomes and metrics associated with traditional scientific approaches, thereby offering much deeper fundamental understandings that can ultimately transform humanity and improve our world.

In exploring and considering the validity of findings in Transcendent Physics, it's interesting to note that many physicists entertain elusive theories such as the Many Worlds Theory; Bosonic String Theory, with its 26-dimensional space-time, Superstring Theory, with its 10-dimensional space-time; and M-Theory, with its 11-dimensional space-time—all of which may be ultimately experimentally unprovable as viable theories.

Perhaps there's a more fundamental fine point. During the 16th century, Galileo made a strong push to separate the "soft" science of experiential sensory signals—vision, taste, hearing, touch, smell, equilibrium, and thought—from signals that fall within "hard" science and can be accurately and quantitatively described and modeled by mathematical physics—for example, the line spectra of the hydrogen atom or the diameter of the heliocentric orbit of the planet Jupiter.

Philosopher Philip Goff has referred to this as *Galileo's Error*.[3] This may well be why a quantitative explanation of some of the more "metaphysical" phenomena of quantum mechanics, such as the Observer Effect, teleportation, and quantum entanglement, have not been accessible by the current stage of development of that science.

Since the 18th century, the Scientific Method has represented the backbone of credible modern research. The process typically involves gathering existing knowledge regarding the issues at hand; formulating a hypothesis; conducting experiments to test the hypothesis; analyzing the output data; developing conclusions; projecting outcomes for future experiments; and continuing this cycle iteratively, until consistent, correct predictions arise for the results of new experiments.

Due to the paradoxically simple yet intrinsically complex nature of consciousness, this traditional approach is inadequate, and often not possible. Current instruments can measure correlates of consciousness but are incapable of isolating and quantitatively analyzing their true fundamental nature. In a very real sense, we are asking, What are the specific causal metrics or parameters that create a given thought or experience? And if we were to interject those parameters into a conscious individual, could we quantitatively, with great accuracy, reproduce those same experiences? This line of thought follows from the definition of *consciousness* by philosopher Thomas Nagel: "What is it like to be a bat?"[4]

Intriguingly, the most effective tool for studying this intrinsic essence of the universe is human consciousness itself, particularly at elevated energy levels. Under such conditions, it can access information unobtainable by any known physical apparatus. Often categorized as *pseudoscience*, this is an outcome that is very difficult for even the brightest and most experienced physicists to accept. It requires a shift in thinking that challenges the very core of the Scientific Method.

Currently, human consciousness can only be explored in qualitative studies. Discovering a means to move to quantitative research is needed and would be a major breakthrough. Just as quantum mechanics in its evolution to *Conscious Quantum Mechanics* (in Part II of this book) has contributed to our general understanding of the correlates of consciousness, I believe that neuroscience will continue to elaborate on the biological mechanisms of these correlates. However, I seriously doubt that it can go any further in understanding the fundamental essence of consciousness. Hence, my bet is on Transcendent Physics, assisted and supported by advances in neuroscience.

The second reason for the dearth of rigorous scientific consciousness research is the prevailing belief among many scientists that exploring consciousness is merely the illusion of misguided scientists such as transcendent physicists. The former group of scientists maintains that understanding consciousness should solely adhere to the Scientific Method, fearing that any deviation toward what they perceive as the "metaphysical" methods of Transcendent Physics would jeopardize their professional reputation and standing in the scientific community. Such is the allure of money, recognition, professional advancement, and ego satisfaction.

Max Planck experienced similar skepticism when he introduced his concept of photons as quanta of energy, and Galileo endured a lifetime of house arrest for proposing heliocentrism. This is often the legacy of advanced innovative thinkers who confront significant resistance when challenging the status quo, which is usually tied to financing, and the possibilities for professional advancement.

Not far behind these factors is the egocentric crisis precipitated by taking an abrupt turn away from what is perceived as high-quality scientific research into uncertain areas, and worse yet, into the unacceptable territory of what is perceived by most as pseudoscience.

This book is not a textbook. Rather, it aims to provide a detailed primer on Transcendent Physics, focusing on the foundational principles of this discipline.[5] It is intended to be of interest to both the novice and expert, who are inquisitive about the nature of consciousness and its immense possibilities. Drawing a parallel with the transition from Galileo's theories to Newtonian physics, and subsequently to quantum physics, we will introduce relevant basic elements of what could be termed *Conscious Quantum Physics,* a stepping stone to Transcendent Physics.

In Part II, *A Primer on Conscious Quantum Mechanics*, I have tried to clearly and accurately translate complex concepts from mathematics and quantum physics into an acceptable vernacular. For readers not having a basic exposure to quantum physics, it may be a bit more challenging than the rest of the book; however, it is definitely worth your effort, as it presents important subtleties that pervade our universe and provides valuable insight into the fundamental ethos of consciousness, the workings of the cosmos, and the very nature of *True Reality*—your reality.

A final important point: This book consists of *24 Reflections*—not 24 instructive *chapters*. The purpose of this structure and nomenclature is to emphasize that the content is not, in many cases, based on proven truths, but on my personal reflections and Conjecturisms over decades of study, contemplation, and meditation. I have drawn from the works of numerous recognized scientists, philosophers and, especially, Wisdom Seekers,[6,7,8,9] who, during the last few millennia, studied and have written about the nature of reality, consciousness, and the universe. In a sense, I have merged what I feel are the most relevant teachings of Eastern and Western philosophers with the science of conscious quantum physics.

My intent is not to convince you of my conclusions, but to challenge and stimulate your thinking on the nature of consciousness and encourage further debate and objective study to uncover the true essence of the only intrinsic element of our universe.

I've had a similar objective in the fiction I've written. For example, in the Luc Ponti Chronicles Trilogy, consciousness plays a critical role throughout all three novels.[10,11,12] I was pleased to see that bestselling author Dan Brown has recently released his latest novel using a similar approach in *The Secret of Secrets*,[13] which takes place in Prague, the magical city I call home. Quantum physics, metaphysics, and consciousness play critical roles throughout this novel.

Understanding the essence of consciousness is important to everyone because it offers vital possibilities for humanity's future. This inquiry deserves serious scientific consideration and exploration, free from the constraints of fear, and endowed with the freedom and flexibility to pierce the long-standing veil of exclusivity for the Scientific Method. This could well pave the way to a sustainable path forward and a "quantum leap" in our understanding of True Reality, our own purpose, and that of the entire universe.

PART I

The Case for Transcendent Physics

"We see the world piece by piece, as the sun,
the moon, the animal, the tree; but the whole,
of which these are shining parts, is the soul."

—Ralph Waldo Emerson

REFLECTION 1

From Alchemy to Transcendent Physics: A Historical Perspective

Evolution of *Scientific Inquiry* to the Frontier of Consciousness

"I'm not interested in preserving the status quo.
I want to overthrow it."

—*The Prince* (1532), Niccolo Machiavelli

Introduction

The trajectory of scientific thought, spanning centuries, is a testament to humanity's restless curiosity and its relentless pursuit of truth. Beginning in the shadowy workshops of alchemists and the speculative treatises of natural philosophers, the methods and aims of science have undergone radical transformations, creating enormous value for humanity. Each epoch has brought a new lens through which to perceive reality, often propelled by both necessity and the brilliance, creativity, and courage of a few singular minds.

Today, as we grapple with the enigma of consciousness and its relationship to the physical world, we stand on the brink of yet another paradigm shift—a movement from the solid ground of quantum physics toward the as-yet-unmapped territory of what some have begun to call Transcendent Physics.[14]

From Alchemy to the Scientific Method

In the 16th century, the prevailing scientific approach to understanding nature was deeply entwined with *alchemy*. With its roots in ancient Egyptian and Greco-Roman traditions, alchemy was as much a spiritual practice as it was protochemistry. Many so-called alchemists sought to transform base metals into gold and concoct elixirs of immortality. But the true alchemists—perhaps few in number—like the Wisdom Seekers of the East, sought to use their powers to connect with *Universal Consciousness* and unravel the secret correspondences between the macrocosmos and the microcosmos.

At the time, the methods of most alchemists comprised a patchwork of tradition, intuition, and experiment, with minimal formal structure. The processes were subjective, often secretive, and not always reproducible. Alchemical knowledge was guarded, allegorical, and communicated through arcane symbols. Despite its mystical veneer, alchemy did plant the seeds of experimental inquiry; alchemists' meticulous documentation of processes and results laid the groundwork for later systematic investigations.

The 17th century, however, ushered in a revolution. Figures such as Francis Bacon articulated the principles of the Scientific Method: observation, hypothesis, experiment, analysis, and predictive future results. This methodological shift marked a decisive break from alchemy's esoteric traditions. Inquiry became public, collaborative, and above all, repeatable. Nature was no longer a veiled mystery to be divined but a system to be interrogated, measured, and explained.

The Age of Physics: Galileo to Newton

With the dawn of this Scientific Method, the study of physical phenomena was transformed. Galileo Galilei, often called the "father of modern science," pioneered the use of mathematical models and empirical tests. His observations of celestial bodies through the newly invented telescope, and his experiments with falling objects, undermined centuries-old Aristotelian dogma.

Isaac Newton, building upon Galileo's foundation, synthesized the laws of motion and universal gravitation. With Newton's "Principia Mathematica" (1687), the universe became a vast, deterministic mechanism—clockwork governed by precise mathematical laws. Newtonian physics, or classical mechanics, provided the scaffolding for nearly all subsequent technological and scientific advances for more than two centuries.

This era was characterized by a confidence in the power of reason and accurate measurement. The boundaries of reality were thought to be penetrable by human intellect; and the universe was, in principle, considered fully knowable.

The Quantum Revolution: Shattering Certainty

By the late 19th and early 20th centuries, cracks began to appear in the edifice of classical physics. Phenomena at the atomic and subatomic scales could not be reconciled with Newtonian mechanics. Enter quantum physics—a discipline inaugurated by Max Planck's discovery of energy quanta; and transformed by Albert Einstein, Niels Bohr, Werner Heisenberg, and Ewin Schrödinger.

Quantum mechanics revealed a world that defied classical intuition. Particles behaved as both waves and particles, existing in a superposition of states until measured. Reality, it seemed, was not deterministic but probabilistic—subject to the uncertainties embodied in Heisenberg's Uncertainty Principle. The observer could no longer be regarded as a passive spectator; their measurements influenced the outcomes. This strange new paradigm forced physicists to confront the limits of objectivity and the possible role of consciousness in collapsing quantum states.

Consciousness and the Limits of Physics

Despite its predictive power and its centrality to modern technologies, quantum physics struggles to address the most profound of all mysteries:

consciousness. How does subjective awareness arise from the interplay of physical particles? Is mind a property emerging from matter, or is it a fundamental aspect of reality?

Attempts to answer these questions have taken many forms: Some, like the Copenhagen Interpretation, accept the limitations of our knowledge; others, such as the many-worlds hypothesis, posit infinite branching universes to account for quantum indeterminacy. Yet, none have fully unlocked the riddle of subjective experience.

As neurobiology and cognitive science map the architecture of the brain, the "Hard Problem"—the question of why and how qualia (subjective experiences) arise—remains unresolved. Here, the rigorous reductionism of physics meets the intractable mystery of the *self*.

Toward Transcendent Physics: The Next Step

To fully explore and perhaps harness the potential of consciousness, a growing chorus of philosophers and scientists is calling for a transition beyond quantum physics—a leap into what some describe as Transcendent Physics.[15] Currently, much of this emerging field is speculative, but it draws upon quantum physics, the philosophical insights of both Eastern and Western Wisdom, the findings of parapsychology and metaphysics, and unresolved anomalies of current scientific paradigms.

Based on findings in the abovementioned fields, we can ask, What might define Transcendent Physics?

- **Integration of Subjectivity:** Rather than relegating consciousness to an epiphenomenon, Transcendent Physics might treat it as an *intrinsic* aspect of reality, intertwined with the fabric of the cosmos.
- **Nonlocality and Interconnectedness:** Extending the quantum notion of quantum entanglement, this new area of physics may explore the possibility that consciousness itself is nonlocal, distributed, and even collective.

- **Multidimensional Frameworks:** Beyond the familiar dimensions of space and time, Transcendent Physics could propose realms or planes where consciousness operates, potentially accessible through altered states or new, advanced technologies.
- **Participatory Reality:** The observer is not only a measurer but potentially a co-creator of reality, blurring the lines between subject and object.

These are, for now, the contours of a vision rather than a consensus. The transition to Transcendent Physics will require new mathematics, novel experiments, and a willingness to question the very foundations of the scientific enterprise, and more fundamentally, the Scientific Method.

Challenges and Opportunities

This bold frontier is not without its critics. Many scientists caution against the dangers of pseudoscience and the abandonment of empirical rigor. Yet, history reminds us that each revolution in thought—from alchemy to quantum physics—was once derided as impossible or unscientific.

The pursuit of Transcendent Physics invites us to revisit ancient questions with new eyes. It suggests that the next great leap in science may not be *outward*, into the reaches of space, but *inward*, into the depths of the mind—and more specifically, the *soul* or *true self*. If we are to understand and access the full potential of consciousness, we must be prepared to rethink not just our theories, but the very nature of inquiry itself.

Conclusion

The journey of science is a journey of transformation—of our methods, our understanding, and our sense of possibility. From the cryptic laboratories of alchemy to the elegant equations of Newton and the paradoxes of quantum mechanics, we have repeatedly reinvented our relationship with the universe. Now, as we confront the mystery of consciousness, science stands before a new threshold. The move from quantum to Transcendent Physics promises not only fresh insights into the cosmos, but perhaps a deeper understanding of ourselves. What shape it will ultimately take remains to be seen. Yet, as always, the adventure of discovery calls us onward.

REFLECTION 2

Cosmic vs. Conscious[16] Cosmology

"In the beginning, there were only probabilities. The universe could only come into existence if someone observed it. It doesn't matter that the observer turned up several billion years later. The universe exists because we are aware of it."

—Lord Martin Rees, astronomer royal

In exploring modern cosmology, evolutionary theory, geological records, molecular genetics, and epigenetics, we encounter two broad schools of thought regarding the nature of True Reality. The first, Cosmic Cosmology (or Physical Cosmology), posits that everything that has occurred since the dawn of time is a series of accidents or coincidences, including your existence and that of our planet—a minute speck in the vast space-time of cosmic architecture.

In contrast, the second school of thought, Conscious Cosmology,[17] suggests that our sensory experiences offer only fragments of a larger reality. What Cosmic Cosmology labels as accidents or coincidences, Conscious Cosmology asserts is a synchronization and correlation among all elements of the universe, implying that there are no true accidents.

Universal Connectivity

The universal connectivity of all energy and matter finds its roots in one of the greatest cosmic mysteries stemming from the birth of our universe, commonly described by the Big Bang Theory. This theory posits that all energy and matter originated from an unimaginable explosive expansion of an unfathomably small singularity[18]—smaller than an electron—emerging within an infinite realm of "nothingness." *Within that singularity existed "something," identified by some philosophers and scientists as pure consciousness, the Unified Field,* or a *Divine Source*; the latter, because of its unique capabilities, is referred to by some as *God.*

Consciousness Is Intrinsic to the Universe

Since the advent of quantum physics, many scientists have grappled with what they consider the mystical aspects of consciousness. Max Planck, who discovered quantum mechanics, articulated this in his 1931 interview with the *London Observer*: "I regard consciousness as fundamental. Matter and energy are derived from consciousness. We cannot go beyond consciousness. Everything we talk about, everything that exists, postulates consciousness."[19] Planck, along with many of his contemporaries, concluded that consciousness is *intrinsic* to the universe. Unlike matter, which can be endlessly subdivided, consciousness is indivisible—it is the essence of the universe and of all life, matter, and energy.

For our purposes, we define *consciousness* as a subtle, transcendent energy that provides awareness of, and responses to, one's inner self; the environment of our four-dimensional world; and all consciousness that exists throughout the cosmos and beyond. It is directly or indirectly responsible for all agency that occurs throughout the universe.

These subtle energies exist as a broad spectrum, forming connections among all observable energy and matter throughout the universe, and likely beyond as well. Paradoxically, while these consciousness energies are much more subtle than classical or quantum energies—such as ultraviolet, infrared, gamma, or even cosmic radiation—they possess far greater transformative potential in our four-dimensional universe.

Understanding Conscious Cosmology requires principles from quantum mechanics, but even more critically, it necessitates deep insights from the emerging field of Transcendent Physics.[20] Just as our current understanding of the universe evolved from the progressive transition of Galilean to Newtonian to quantum principles, a true comprehension of consciousness and its role within the universe requires the framework of Transcendent Physics.

The True Ground-of-Being

Conscious Cosmology asserts that consciousness is the true *ground-of-being*—not physical form. The consciousness that generates thought is the same consciousness behind the intelligent activities of the universe. This framework expresses itself through a *triad* of consciousness:

1. **Immanent Consciousness** is the pervasive awareness interwoven within every energy field and all particles of matter in the universe. This is True Reality, which is infinite, nonlocal, and eternal. Often referred to as *spirit* or *soul*, Immanent Consciousness transcends the confining geometry of the human body and a lifetime. It is nonlocal, present throughout the entire universe, and infinite in time.

2. **Collective Consciousness** is the interconnected web of all beings and all physical and nonphysical entities, creating unifying connectivity among all energy and matter throughout the universe. Collective Consciousness represents the interdependence of all energy and matter, affirming that our spirit is not localized; it is everywhere, always has been, and always will be.[21] As Francis Thompson incisively metaphorized in his poem "The Mistress of Vision": "Thou canst not stir a flower without troubling a star. . . ."[22]

   While Collective Consciousness resides in the physical world at the interface of all energy and matter, high-energy levels of Immanent Consciousness exist in the gap between thoughts during meditation. *Herein lies the foundation of all creativity, the place where you can go to access the power of pure consciousness to manifest what you need or desire.*

3. **Cosmic Consciousness** is an infinite, eternal, *Divine*[23] Source resonating with *Six Cosmic Core Values*: truth, wisdom, understanding, forgiveness, love, and compassion. These values are not arbitrary or based on any kind of moral injunction, but as noted in the next section, are essential for unfolding the *Meaning and Purpose of the Universe.*

   Each of us participates in the creative process of being within our universe, contributing to Cosmic Consciousness—an eternal Divine Source without beginning or end. These comments are spiritual in content, but have no connection with any religion—although Cosmic Consciousness may be seen by some as equivalent to what they perceive as *God.*

The essence of all three functions of consciousness is exactly the same. The only important difference is the energy levels, with Cosmic Consciousness having the highest level, perhaps infinitely high.

Meaning and Purpose of the Universe

The Meaning of the Universe embodies the continuous harmonious integration of physical evolution with consciousness evolution throughout the cosmos. Importantly, *physical evolution relies on consciousness, and the evolution of consciousness cannot occur without physical processes; they are inextricably intertwined.*

The Purpose of the Universe *is to perpetually elevate the energy level of its overall consciousness.* As emphasized above, all consciousness is identical; its only variable is energy level. Consciousness is propagated throughout the universe by an energy spectrum unrelated to the electromagnetic spectrum. For this reason, its speed of travel is not limited by Einstein's Special Theory of Relativity. It travels instantaneously.

Similar to the classical conservation laws for energy and matter, consciousness can be neither created nor destroyed, only transformed into energy or matter. Conceptually, as Einstein demonstrated the reversible interconvertibility of energy and matter, this holds true for consciousness, as well: Consciousness $\rightleftharpoons$ Energy or Matter.

Cosmic Cycles

The implication of these transformations combined with the Big Bang Theory are the bases for infinite *Cosmic Cycles.* Namely, consciousness creates energy and mass, which over long geological time frames—many billions of years—expand, ultimately contract, and then revert back to consciousness—ad infinitum. At the end of each contraction, a new singularity forms, containing nothing but pure consciousness from the prior universe, ready to expand and form the next universe. It is completely *informed,* as it contains the totality of information of all prior universes.

In accordance with the Purpose of the Universe, each of these cycles elevates the total energy level of consciousness throughout the universe—*Universal Consciousness*—not the amount which is constant.[24] Some cosmologists refer to this model for universe formation, expansion, collapse, and subsequent reformation as the *Big Bounce.*[25,26,27]

An important postscript to this model as applied to planet Earth is the uniqueness of the human species. As our planetary physical and consciousness evolution have proceeded, at this point in time, human beings are at the pinnacle of both evolutionary processes. Within this focus, one might inquire, "What, then, is the Meaning and Purpose of Life?"

Meaning and Purpose of Life

The Meaning of Life is for each species to discover, embrace, and further develop its innate attributes. Each of us, as well as all other species on Earth, come into this life with certain specific talents or capabilities. *The Purpose of Life is to use these attributes to contribute to Earth's—and therefore, the universe's—physical and consciousness evolution.* On Earth, this enhances our world by fostering a globally favorable, sustainable, and more aware existence for all.

Putting all of the above elements together, it can be seen that our universe is *nondualistic,* meaning that Immanent, Collective, and Cosmic Consciousness permeate all energy and mass throughout the universe—

and highly likely beyond all space-time as well. All is connected by consciousness—that is, *all is One*.

This is in contrast to a *Dualistic Universe* model, which maintains that nothing is connected. As we shall see in later Reflections, nondualism provides us with the possibility of great capabilities, provided we elevate our consciousness to high energy levels by quieting the fear, desire, and chaos of our four-dimensional, diversional, material world. Dualism, by maintaining that nothing is connected, is the cause of our most perilous challenges and problems.

The Power of Nonduality

When comparing these two models of the universe, it's important to note two relevant states-of-being. *Transcendence* is a state-of-being beyond our physical world—that is, beyond space-time; whereas *Immanence* is a state-of-being within our physical world. To say that the *Divine (God)* is *Transcendent* means that *God* (*Cosmic Consciousness*) is beyond us in space and time. To say that *God* is *Immanent* means that *God* is an intimate part of us within the physical universe.

Most religions emphasize Transcendence—that is, *duality*: "We are down here, and the Divine is up there." However, the nondualistic model, while embracing Transcendence, emphasizes Immanence—that is, *nonduality*, meaning that "the Divine is part of us." And thus, the Divine is both Transcendent and Immanent—within and without, as above, so below—providing us with the possibility of great capabilities.

These capabilities can be accessed when we are at high levels of consciousness and aligned with the Six Cosmic Core Values. In this state, we can bring forth anything into our physical world, as long as it adheres to the two *Laws of Manifestation*:

- **Nonviolence:** It will cause no harm by adhering to sociocultural ethical norms governing human and nonhuman life.
- **The Greater Good:** It will further the world, however modestly, toward improved conditions reflecting the Six Cosmic Core Values.

Thus, through elevated states of consciousness—for example, deep-state meditation—to lower the fear and chaos of our physical world, and abiding by the two Laws of Manifestation, we have the ability to change ourselves for the better as well as the world around us.

REFLECTION 3

Which Body Are You?

"The perception of reality is merely an illusion,
albeit a very persistent one."

—Albert Einstein

Three Paths to Reality

What is reality? We can explore this important question through three distinct lenses: our senses, our minds, and our consciousness.

The first approach involves our sensory perceptions. To investigate the surface structure of the moon, we rely on telescopes to extend our vision. Similarly, to analyze the shape and structure of microbes, we utilize high-powered microscopes to amplify our visual acuity. Yet, this sensory exploration often leads to a limited and distorted understanding of reality.

The second way to understand reality is through the eyes of the mind—the intellect. Grasping concepts like the Pythagorean Theorem requires a familiarity with Euclidean geometry, a framework constructed within our cerebral intellect. Understanding Einstein's Theory of Relativity necessitates engaging with the thought experiments that shaped the minds of visionary scientists. Quantum theory, too, demands an appreciation of the intricate mathematics and physical principles conceived by brilliant changemakers in physics such as Erwin Schrödinger, Werner Heisenberg, Paul Dirac, and Max Planck.

However, within the scientific community, almost from the very beginning of science, a tension emerged regarding the relationship between quantum mechanics and consciousness—a connection often sidestepped in academic discourse. Max Planck emphasized this link when he stated that all energy and matter stem from consciousness, the very source and force behind the Big Bang. This insight challenges the intellectual paradigms held by many foundational physicists.

Quantum mechanics stands as one of the most successful frameworks in science, responsible for approximately 40 percent of our global GDP. Technology such as email, smartphones, computers, the internet, and laser surgery all rely on principles of quantum science, which asserts that the material world is, at its core, nonmaterial. Everything we refer to as "stuff" is, in fact, "nonstuff."

This understanding arises from realizing that atoms—the fundamental units of what we classify as physical matter—are not solid entities. Traditional views, like those of Isaac Newton, depict atoms as solid balls, yet quantum mechanics reveals them as hierarchies of information and energy existing within a vast emptiness—more on this in later Reflections. This conclusion emerges when we examine the universe through the introspective lens of consciousness—the intuitive spirit of True Reality.

This brings us to the third approach to understanding reality: perceiving through the lens of consciousness itself—which some might refer to as the *spirit* or the *soul*. An incisive expression of this perspective is found in William Blake's poem "Auguries of Innocence":

We are led to Believe a Lie
When we see not Thro' the Eye
Which was Born in a Night to perish in a Night
When the Soul Slept in Beams of Light.

Blake articulates an important truth echoed by contemporary philosophers and scientists: Our sensory experiences can deceive us. Wisdom Seekers have long comprehended this principle, advising us not to confuse our sensory interactions with the essence of consciousness. Buddha succinctly stated around 500 BCE: "If you can see it, if you can touch it, if you can taste it, if you can smell it, if it is solid, if you can think about it, if you can conceptualize it, if you can visualize it, then—it is not real."

This assertion underscores that the reality perceived through the awareness of our senses is merely a projection of a deeper, unseen consciousness. Without consciousness, we cannot imagine, introspect, think, or perceive. As quantum physicist David Bohm remarked, "What we perceive through the senses as empty space . . . is the ground for the existence of everything, including ourselves. The things that appear to our senses are derivative forms and their true meaning can be seen only when we consider the plenum in which they are generated and sustained, and into which they must ultimately vanish."[28]

Our senses may suggest that the world is flat, yet we now know otherwise. They may lead us to believe the ground is immobile, while the Earth spins on its axis at 1,000 miles per hour, hurtling through space on its orbit around the sun at 67,000 miles per hour. They may present our bodies as static structures, separated by space and time, but in truth, these perceptions are grand illusions.

Conscious Cosmology Aligns with Quantum Mechanics

According to Conscious Cosmology, which aligns with quantum mechanics, *the universe is an interdependent confluence of events existing across past, present, and future within a boundless field of consciousness that has neither a beginning nor an end.* Our senses, deceptive and fleeting, provide only a superficial understanding; when we look at one another, we see fixed anatomical forms, yet the foundation of all physicality—the atom—is itself nonphysical, comprising dynamic streams of energy and information in perpetual exchange with the universe.

You might wonder, *How can I pick up this book and perceive it as solid? Why does the ground feel solid beneath my feet?* This sense of solidity arises from electrostatic repulsion of electrons in the atoms of your feet with electrons in the atoms of the ground, but is more fundamentally explained by a quantum principle known as the *Pauli Exclusion Principle.* This principle states that two or more identical particles, such as electrons with half-integer spin quantum numbers,[29] cannot simultaneously occupy the same quantum state or physical space within a quantum system.

In simpler terms, things that appear to be solid are made up of atoms, which contain electrons. The electrons exist in various quantum states. Those atoms with electrons having the same spin quantum number are not permitted to occupy the same space. The reason you don't fall through the Earth to the other side is because atoms in your feet have electrons with a spin quantum number equal to those within the atomic structure of the ground you stand on. In short, your feet and the terra firma are not permitted to occupy the same space—hence, your perception of solidity. It's that simple!

REFLECTION 4

Your Physical Body

"The human body is a machine which winds its own springs."

—*L'Homme Machine* by Julien Offray de la Mettrie[30]

Sharing Our Atoms

The body you began this journey with is not the same one that exists as you read this passage. With each breath, you inhale approximately 100 billion trillion ($1x10^{22}$) atoms from the physical universe and exhale about the same number. These atoms are sourced from every cell of your body. At the atomic level, you are literally releasing fragments of your liver, heart, brain tissue, and more. With every inhalation and exhalation, you are sharing bits and pieces of your very selves with others. As Walt Whitman eloquently expressed in "Song of Myself" "For every atom belonging to me as good belongs to you."[31]

This is no longer merely a metaphor of poetry; it is a scientific fact of biology. If we envision a spherical boundary around the Earth's atmosphere, recognizing the dynamic equilibrium and exchange of atoms and molecules among all living organisms, we can ascertain and calculate that since atoms and molecules don't disappear, as you read these words, your body contains at least one million atoms that once emanated from the body of historical figures such as Jesus Christ, Buddha, Genghis Khan, Elvis Presley, Madam Curie, or anyone else you can conceive of.

Due to this constant interchange of atoms and molecules in our atmosphere across the globe, over the past three weeks alone, a quadrillion ($1x10^{15}$) atoms have traversed your body, having also cycled through every other living species on the planet. Envision a camel in Saudi Arabia, a taxi driver in Kolkata, or a pigeon in China—at this very moment, atoms within you were once part of those bodies or environments just weeks ago.

You're Never the Same

In less than a year, you effectively replace more than 98 percent of all atoms in your body, exchanging them with the global environment. At the atomic level, you recycle your liver every six weeks, your skin monthly, your stomach lining every five days, and your skeleton every three months. Even your DNA, which encodes the collective memories of millions of years of evolution, experiences this cycle of exchange every six weeks. Like migratory birds, its atoms come and go.

Thus, if you identify solely with your physical body, you encounter a perplexing challenge: Which version of your body are you referring to—the one that commenced reading this sentence or the one existing in this moment?

I began writing this book in 2024, yet my body has transitioned into a 2025 model. The 2024 version is gone, having been composed of elements from various life forms on the planet, including minute particles of atmospheric dust. It circulated within what I refer to as "myself" before returning to its origin—now mingling within other life forms and particles of dust—*"dust to dust."*

Consciousness Transcends Death

However, as consciousness—encompassing memories, dreams, imagination, inspiration, intuition, insight, creativity, and decision-making—we transcend the "death" of the atoms and molecules forming the physical bodies through which we express ourselves. In reality, we must not define ourselves solely by our physical bodies!

"We are such stuff as dreams are made on."

—*The Tempest*, William Shakespeare

If you were able to perceive the physical world as it truly exists—not filtered through the deceptions of your sensory experiences, which often mislead—you would encounter a radically different universe.

REFLECTION 5

Sensory Lies

"The five senses are the ministers of the Soul."

—Leonardo da Vinci

Sensory Deception

Our senses deceive us in a bizarre manner. The brain, responsible for our perception of the world, has never directly experienced the world. How could it? Since birth, it has been confined within the skull, responding only to internal biological signals—changes in pH, electrolyte concentrations, hormones, and body temperature. This biological activity ultimately translates into a binary code of electrical charges across neuronal membranes, forming the foundation of our experiences. But how do these electrical signals generate the rich tapestry of sight, sound, taste, smell, touch, and thought? It remains one of the greatest scientific mysteries of our time.

"The best explanation we have is that something unknown is doing we don't know what!"

—Sir Arthur Eddington, British cosmologist (1882–1944)

Even leading scientists who have studied perception extensively, such as the late Nobel laureate Sir John Carew Eccles, recognized the illusory nature of sensory experience. He famously stated, "There are no colors in the real world. There are no textures in the real world. There are no fragrances in the real world. What actually exists out there is some radically ambiguous and ceaselessly flowing quantum soup."[32] The magic lies in our consciousness, which transforms this abstract quantum field into a tangible reality. Out of a sea of quantum probabilities, we conjure up a world, a universe, and a physical self.

The Hard Problem

Consider a simple thought experiment: Picture a sunset over the ocean. I'm sure you can vividly see it in your mind's eye, but where is that image? If we examined your brain, we would find no sunset—only a cascade of electrical impulses and neurotransmitter exchanges. *How does this activity generate the immersive personal experience or qualia of color, warmth, and awe?* Equally important, how does this neural dance construct the entirety of your reality, including your perception of yourself? As we will see in later Reflections, this mystery is known to philosophers as the Hard Problem.

The term *Hard Problem* was coined in a 1995 paper, Facing Up to the Problem of Consciousness, by Australian philosopher and cognitive scientist David Chalmers.[33] He used this label to refer to the significant challenge of explaining how individual subjective experiences arise from physical processes. Chalmers distinguished this challenge from what he termed "Easy Problems"—namely, explaining the specific mechanisms of attention, memory, and the integration of relevant information, all of which are measurable and observable. The Hard Problem, on the other hand, specifically addresses the subjective, qualitative aspects of consciousness—like the feeling of what it's like to see green, taste vanilla ice cream, or listen to Dvořák's Symphony No. 9.

In essence, the Hard Problem is the challenge of understanding how physical processes in the brain give rise to the subjective, first-person experience of consciousness. To date, all research efforts to achieve this goal have yet to be successful.

The Illusion of Continuity

Despite the seamless flow of experience, reality is fundamentally discontinuous. Modern physics tells us that what we perceive as solid matter consists of molecules, which are composed of atoms, which in turn consist of subatomic particles—electrons, protons, and neutrons—each vibrating in and out of an infinite void. At the deepest level, even these subatomic particles are manifestations of quantum fluctuations, appearing and disappearing at unimaginable speeds. If we could see the world as it truly is, it would become apparent that both we and the chair you may be sitting on are proportionally as empty as intergalactic space. In fact, 99.9999999999996 percent of what we call matter is void, and the remaining 0.0000000000004 percent is itself mostly empty space.

The millennia-old Buddhist concept of *Śūnyatā*—the "great emptiness"—parallels this modern scientific perspective. Wisdom thinkers such as Patanjali and the Buddha, akin to today's cosmologists, arrived at these insights not through instrumentation but through deep contemplation and meditation. Remarkably, their conclusions resonate quite accurately with modern quantum physics: The world we experience is an illusion crafted by our consciousness.

The Nature of Nothingness

What is this "great emptiness" from which we all emerge? Is it merely a cosmic void; or is it the fertile ground of creation, birthing galaxies, nebulae, rainforests, human bodies, and even thoughts? What *is* a thought? Where does it come from, and subsequently, where does it go? These questions have been pondered throughout history. Some claim to have found the answers, but as the saying goes, there's a fine line between sages, geniuses, and psychotics.

> *"The popular mythology of creative genius depends on beloved stereotypes of the artist in youth and old age: the misunderstood upstart who forces us to see the world afresh; and the revered sage who shows us depths of insight attainable only through a lifetime of hard-won experience."*
>
> —Martin Myles Filler,[34] architecture critic

Discontinuity of Reality

A neon sign appears to emit continuous light, but in reality, individual bulbs are switching on and off in a precise timing sequence. Our eyes perceive only the illuminated moments, not the intervals of darkness, creating the illusion of movement. Similarly, a film consists of individual frames moving through a projector at 24 frames per second, yet we perceive smooth motion. This same principle applies to the fundamental nature of reality—at the quantum level, everything is flickering between existence and non-existence at incomprehensible speeds.

Max Planck's groundbreaking discovery in 1900 of the quantum of energy was born out of frustration. Conventional physics insisted that energy flowed in a continuous, uninterrupted stream, yet Planck found that only by assuming that energy was quantized—transmitted in discrete individual packets called *quanta*—could he explain blackbody radiation.[35] This radical insight, initially disparaged by other scientists at the time as nonsense—gave birth to quantum mechanics, forever altering our understanding of reality.

If everything we perceive is a series of discontinuous flickers—on and off—what exists in the "off" state? If there is truly nothing between these states, how does reality transition from one state to the next? This question lies at the heart of Transcendent Physics and the enigma of consciousness itself.

REFLECTION 6

The Discontinuity

"The difference between fiction and reality? Fiction has to make sense."[36]

—Tom Clancy, American novelist

Signal On

Scientists have a good idea as to what's in the "on" part of the signal. It's energy and information. For example, if I take a picture with my cell phone and then email it to somebody in Tokyo, what goes from here to there is not a picture but a vibrating signal of small quantum packets of light—photons—namely, electromagnetic energy. Then the picture is re-created, first in the receiver's cell phone or computer, and then in their consciousness when they view the digitally reconstructed image.

Signal Off

So the mystery is not what's in the "on" part of the discontinuity, it's what's in the "off" part. If you ask scientists at the forefront of quantum mechanics research about this, they will tell you something like the following: "There is no energy, no information, no space-time, and no objects."

So, then, what's there? Their best answer is: "There are infinite possibilities." The off part of the discontinuity is a field of possibility waves—what Nobel laureate Werner Heisenberg, while developing his quantum matrix mechanics and the Uncertainty Principle, called the *realm of potentia* in reference to the concept of potentia advanced by Aristotelian philosophy.[37] This recognizes the immeasurable potential for all that was, all that is, and all that will or could ever be. The details of this almost incomprehensible concept and other aspects of the discontinuity are discussed in more detail in Part II. This aligns with Reflection 9, *Conjecturism 1, The Birth of Our Universe*; and *Conjecturism 2, The Primacy of Consciousness*, in which the universe manifests from nonmaterial consciousness into the observable existence of all energy and matter.

Nonlocal Correlation

The second point these scientists will agree on with respect to the discontinuity is a phenomenon called *nonlocal correlation*, which in metaphysics is synonymous with omniscience and omnipresence. This means that everything in the universe is synchronized, correlated, in harmony, and coincidental with everything else. This effect is faster than the speed of light moving across space-time, where past, present, future, and distance in space are all interconnected instantly. This occurs even when there appears to be no apparent force field between objects—at least, none that our current state of science can recognize. This principle resonates with Reflection 9, *Conjecturism 3: The Spectrum of Consciousness*, underscoring the instantaneous correlation between all things.

This nonlocal correlation occurs at superluminal velocities but does not violate Einstein's Theory of Special Relativity, which limits the velocities of all mass and energy to less than the speed of light. This limitation applies only to mass and to energies of the electromagnetic spectrum. It does not apply to massless consciousness, which, according to the principles of Transcendent Physics, can travel at an infinite velocity for instantaneous connection. It is consciousness that carries the information.

Buddha called this synchronization and correlation "interdependent co-arising," by which everything co-arises, correlates, harmonizes, and synchronizes with everything else. This nonlocal correlation effect is responsible for what we call *simultaneity* in the universe: a phenomenon that appears throughout nature, especially in biological organisms.

Bodily Synchronicity

Your body, for example, expresses simultaneity and synchronicity. It is made up of some 33 trillion cells, a number that is comparable to the number of stars in many of the galaxies in our universe. Each of these cells measures less than 1/1000 inch (0.0254 millimeters) in diameter and contains within its DNA, instructions that would fill a thousand 600-page books. Each cell performs some 100,000 activities per second, and each cell instantly harmonizes and correlates its activities with every other cell in your body. But when this system doesn't move with its usual clockwork-like precision, you can become ill—in some instances, gravely.

How does your body keep this up every second of every day, every day of every year, year after year, until, on average, after some 80 to 85 years of operation, it just stops? How does a human body think thoughts, play the piano, kill germs, remove toxins, and make babies all at the same time? And all while it tracks the movements of the stars and planets because your biological rhythms are part of the symphony of the entire "uni-verse"—"one verse," "one song." We are all part of one cosmic symphony.

A Proliferation of Uncertainty

The third attribute of the discontinuity is a proliferation of uncertainty, where the laws of nature become uncertain. This means that as we enter the realm of the quantum world or consciousness, the very act of observing or measuring disturbs the system being observed or measured in such a way that makes it impossible to determine a precise value. This uncertainty mirrors and affirms the observer's central role in collapsing possibilities into physical existence.[38]

Furthermore, as we shall see in Part II, in the quantum world it is impossible to determine when or where a quantum event will occur. Nobel laureate Max Born eventually demonstrated that all we can do is calculate the probability of a quantum event occurring. This is in distinct contrast to classical physics, with which we can measure things very precisely. This means that our classical world is deterministic. Given sufficient information about a given mechanical or chemical system, we can calculate its future about when and where certain events will occur. Not so in the quantum world, which is not deterministic but probabilistic.

When Einstein first heard of Werner Heisenberg's Uncertainty Principle, he said, "I don't believe that God would play dice with the universe." Decades later, world-renowned cosmologist and physicist Stephen Hawking quipped: "So Einstein was wrong when he said, 'God does not play dice.' Consideration of the theory of black holes suggests, not only that God plays dice, but that He sometimes throws them where we can't find them."[39]

Quantum Creativity

The fourth principle of the discontinuity is a form of quantum creativity wherein patterns of intelligence, information, and energy move from one expression to a new expression in a quantum jump, without a transitional phase—that is, no presence in the in-between phase. This reinforces the idea that nature advances in quantum leaps rather than smooth, continuous progressions.

The Observer Effect

Finally, the fifth attribute of the discontinuity in quantum mechanics is known as the *Observer Effect.* First uncovered by the early developers of quantum mechanics and described in detail by the late internationally recognized physicist John Archibald Wheeler, the Observer Effect states that the physical universe would not exist if there were not conscious beings looking at it. Quantum physicists describe observation as leading to the collapse of the wave function, which in turn, creates the observed event or item.[40]

The Discontinuity Is Consciousness

From what has been said above, it is evident that the discontinuity is consciousness itself. It is the spirit, and your soul[41] is a part of this. Your soul is not a thing. Your soul is a field of infinite possibilities. Your soul is the omnipresent, omniscient, omnipotent awareness that is right now orchestrating with simultaneity the hundreds of trillions of things happening in your body in this moment and will continue to do so in each of the succeeding moments until the death of your body; and then your soul will continue on with infinitely more possibilities. One might conclude that to maintain health in your body and mind, take care of your soul! This conclusion is discussed in Reflection 9, *Conjecturism 2: The Primacy of Consciousness*, asserting that consciousness is the primary essence from which all reality emerges.

REFLECTION 7

Your Soul and Reality

"Put your ear down close to your soul and listen hard."

—Anne Sexton, Pulitzer Prize–winning American poet and author

Connecting with Your Soul

One of the wonderful consequences of the body-soul relationship is that it is quite easy to get directly in touch with your soul: Sit quietly in a dimly lit room with soft music playing, take several deep breaths, and release each one very slowly. It is best to breathe in for 5 seconds, hold it for 5 seconds, and then slowly exhale over 10 seconds. Repeat this process several times, dismissing all thoughts, until you are in a relaxed state.

Then become aware of who is listening to the music by silently contemplating, "Who is *really* listening to this music?" If you feel a pleasant, still presence, that's your *soul*—your Immanent Consciousness, your very core. It's who you truly are. It is not your mind, which might be asking, "What is this guy talking about? Why in the world am I reading this book?" That's your *mind*, driven by your ego and subconsciousness, and engaged in constant conversation with itself. But this mental process and dialogue—known as *Choiceless Awareness*—occurs in the presence of your Immanent Consciousness—your *soul*, your true *being*.[42] Find solace in the fact that your mind will dissipate when your body dies—but not your *soul*—it is eternal.

In that *being*, a thought comes and then it goes. In that *being*, an emotion arises and then subsides. In that *being*, a perception of the world emerges and then dissolves. In that *being*, your perception of the entire universe appears and then vanishes. In that *being*, the atoms and molecules composing your body come and they go. Once, you had the body of a child—with a different mind, emotions, and personality. In that *being*, the body, the mind, emotions, and perceptions—the entire universe, all come and they go.

Hold on to that presence; it's the only thing that is truly real about you. Everything else is the arising and subsiding of space-time events within that *being*.

"Because it arises, it will also subside."

—Buddha

Your True Reality

This is one of the noble truths that cause suffering—because we mistakenly identify reality with impermanence. We may think, *We better do as much as we can, as soon as we can, and get as much as we can, because it's all over when we die.* However, if we look deeper, we can see *that* in which all things arise and into which all things subside, just like the universe, has no beginning or ending in time.

It is beneficial to be mindful of the arising and subsiding of form, feelings, and thoughts—of "This is me" and "This is mine." By piercing through this veil, you reach the eternal ground of consciousness—the *nondual, nonlocal consciousness* in which all arising and subsiding takes place.

Recall the Big Bounce theory of the universe, where infinite cycles of the universe's formation, expansion, contraction, dissolution, and reformation occur. The universe arises and then subsides. You are an intimate part of the universe; therefore, you, too, arise and subside—both physically and in your Immanent Consciousness, your soul. But just as the universe is reborn, so, too, does your consciousness evolve in infinite cycles, increasing its consciousness energy level with each cycle.

This understanding offers a practical pathway to True Reality—both yours and that of the universe. It opens a window to the realm of infinite consciousness in which we all coexist and co-create. Death is not the end; it is merely another step along our infinite journey. When you experience this consciousness directly—not as a mere belief but as an undeniable reality—you will come to intimately understand the Six Cosmic Core Values of the universe, and ultimately, the intoxication of *love.* Love is the source of all healing, all creativity, and all universally beneficial manifestations. One of the best-kept secrets is that love, which is pure high-energy consciousness, and incompletely defined within the English language, is one of the greatest forces in the universe (see Appendix II). It is consciousness in action.

The Path to True Reality

These insights are the foundation of Buddha's philosophical teachings, articulated in what he called the Four Noble Truths:

1. Every life contains suffering.

2. There are reasons for it.

3. There is a way out.

4. The way out is through the *Eight-Fold Path to Enlightenment.*

The Eight-Fold Path is a guide to spiritual and intellectual enlightenment, aimed at freeing us from attachment to, and delusion about, our physical universe. It entails seeking and achieving the:

1. **Right View:** Embrace the Four Noble Truths and the reality of suffering.

2. **Right Intention:** Cultivate thoughts free from greed, hatred, and cruelty.

3. **Right Speech:** Speak truthfully, kindly, and constructively.

4. **Right Action:** Act ethically, refraining from harmful deeds and misconduct.

5. **Right Livelihood:** Earn a living through ethical means that do not harm others.

6. **Right Effort:** Apply diligent efforts that cultivate positive states of mind.

7. **Right Mindfulness:** Develop continuous cognition of your body, feelings, thoughts, and the world around you.

8. **Right Concentration:** Cultivate a deep meditative awareness to achieve a state of clam and deep insight.

From this path, four important conclusions arise:

1. **The relative is impermanent:** Everything that arises in our four-dimensional world of space and time must also subside. The only exception is consciousness, which is *intrinsic* to the universe and is the only thing that truly exists—eternally. Your *Immanent Consciousness* is your *True Reality*.

2. **The separate self does not exist:** Our souls are mirrors of each other's souls.

3. **Everything is interconnected:** We exist as inter-beings that arise together in the domain of "inter-isness."

4. **Nondual consciousness is the fundamental reality:** In this reality, where all is connected, both the observer and the observed co-arise simultaneously.

Remarkably, these conclusions were rediscovered by quantum physicists some 2,500 years after Buddha's formulation of the same concept. Modern physics now points toward a universe where observation and reality are entangled, where consciousness plays a fundamental role in the very structure of physical existence. This convergence of ancient wisdom and cutting-edge science reaffirms the eternal truth: *Consciousness is the intrinsic foundation of all reality.* This fundamental truth is embraced within the words of Sri Nisargadatta Maharaj, who focused much of his life on the nature and power of consciousness:[43]

> *"The way to truth lies through the destruction of the false. To destroy the false, you must question your most inveterate beliefs. Of these the idea that you are the body is the worst. With the body comes the world, with the world—God, who is supposed to have created the world and thus it starts—fears, religions, prayers, sacrifices, all sorts of systems—all to protect and support the child-man, frightened out of his wits by monsters of his own making. Realize that what you are cannot be born nor die and with the fear gone, all suffering ends. What the mind invents, the mind destroys. But the real is not invented and cannot be destroyed. Hold on to that over which the mind has no power."*

REFLECTION 8

Personal Principles of Transcendent Physics[44]

"In the deeper reality beyond space and time, we may be all members of one reality."

—Sir James Jeans, British physicist (1877–1946)

Everything we have explored so far can be distilled into 15 Personal Principles, any one of which, practiced with dedication and mindfulness, can catalyze genuine personal transformation. Each Principle invites you to draw on the creative power of Cosmic Consciousness from the intersection of modern quantum physics and the timeless wisdom of Eastern philosophy.

A summary of the technical essence of the science—Transcendent Physics—is presented in Reflection 9, with more details in Reflections 22 and 23, where you will find the supporting logic. Here, we will stay *experiential*, providing 15 Principles that support Transcendent Physics and a path to personal fulfillment by contributing to the whole—you, the world, and the cosmos.

A word of orientation: The tone and approach of this reflection is purposefully nonsectarian; yet we borrow freely from the philosophical traditions of Buddhism, yoga, Vedānta, and other Eastern philosophies because, across cultures, they provide the clearest foundations for navigating a path to understanding consciousness. Take what resonates with you, test it in the laboratory of your own mind, and choose that which speaks to you.

Principle 1: Hidden Dimensions of Existence

You inhabit more than flesh and thought. Beneath everyday awareness lies a triad of nested consciousness:

- **Immanent Consciousness:** Your timeless personal consciousness.
- **Collective Consciousness:** A subtle energy field that links all mass and energy throughout the cosmos.
- **Cosmic Consciousness:** The creative Divine Intelligence that permeates all of the universe.

The surface of life is shaped by karma (past action), memory (its imprint), and desire (the impulse to re-create). To pierce that surface and understand the fundamental essence of life, here are a few thoughts for your consideration:

a. Initiate a daily practice of Choiceless Awareness (15 minutes). Sit quietly, breathe as described in Reflection 7, resting your attention in your heart, and enter a state of high creative intuition.

b. Contemplate the six questions that shape your reality. Silently inquire: *Who am I? What gifts are uniquely mine? How may these attributes serve humanity and the planet? What is my purpose? What do I truly long for? What contribution do I want to make to the world?* Don't expect quick answers. Be stoically patient, and live with the questions. Eventually, this practice will create an unmistakable and inspiring feeling from which the answers will flow.

c. Meditate. There are two broad forms of meditation: *Passive meditation* employs a mantra or observing your breath. This calms the mental turbulence that hides your deeper self. The answers are there. *Active meditation* is moment-to-moment present awareness. The "Power of Now," as expressed by Eckhart Tolle, dissolves the ego's constant commentary during ordinary activity. Both meditative routes converge in a quiet bliss that words can barely touch. In that stillness, creativity spikes, solutions appear, and manifestations quicken.

Principle 2: The World Exists within You

You are not in the world; the world is within you, in your consciousness. Quantum mechanics hints at this through the Observer Effect. Upon observation, possibilities and probabilities collapse the wave equation, giving way to a single event. Thus, every tree, every person, every star, is a projection on the screen of your personal awareness. The next time you meet someone, silently note: *This encounter is happening within me.* Notice how your compassion naturally enlarges.

Principle 3: The Path of Yoga

Yoga means "union," from the Sanskrit root *Yuj*, meaning "to join," and all four classic types converge on the same summit:

a. **Bhāva Yoga:** The meditation of being. *Bhāva* is mindful intent, intentionally setting an appropriate mood or mental space within one's self.

b. **Karma Yoga:** Selfless action. A path to reach spiritual liberation through work. It is rightful action without being attached to its fruits, or being manipulated by what the results might be; a dedication to one's duty; and trying one's best while being neutral with respect to rewards or outcomes such as success or failure—a form of pure *stoicism.*

c. **Bhakti Yoga:** Love and devotion. This "path of love" is a spiritual discipline focused on cultivating a relationship with the Cosmic Consciousness that already exists within you, but is often masked to varying degrees by the stress and chaos within our four-dimensional world. This cultivation involves expressing love and devotion through various practices, including meditation, chanting, and service.

d. **Jñāna Yoga:** The yoga of knowledge or wisdom that is an outgrowth of focused thought. Transcendent physics reframes God not as a distant supreme monarch, but as the indivisible subtle energy field of Cosmic Consciousness, which is simultaneously *transcendent* (beyond you) and *immanent* (within you). It emphasizes a path of understanding and self-realization through knowledge, particularly of the true nature of reality and the self.

Practice any one of these four forms of yoga sincerely, and a subtle energy field, which is already within you, begins to become apparent and flow more steadily through you, providing immense clarity as to your purpose and the nature of True Reality.

Principle 4: Fulfillment Comes from Within

Creative manifestation follows a simple algorithm:

- **Intention:** Set a vivid inner image and rationale—the "why."
- **Attention:** Maintain your focus on the overall goal—the "what."
- **Detachment:** Release your thoughts from the detailed mechanics—the "how."
- **Gratitude:** Affirm your current abundance.
- **Trust:** Allow the universe to arrange the details. Keep your goals high level—on the "what." Stay flexible concerning tactics—the "how." Morning and evening meditations (20 to 30 minutes) plug you into the "back office" of the universe, where the details are handled.

Principle 5: Transcending Suffering

Buddha diagnosed five roots of suffering:

- Ignorance of the true self
- Clinging to impermanence
- Fear of loss
- Ego identification—a socially conditioned false self
- Fear of death

Meet each with mindful awareness and the knot unties. Perception then becomes direct, unfiltered by fear and anxiety.

Principle 6: True Freedom Lies in Choiceless Awareness

Between stimulus and response lies a nanosecond of pure witnessing—a gap between thoughts that Indian philosopher Jiddu Krishnamurti called *Choiceless Awareness.* Enter it and you stand at a portal to infinite possibilities. In quantum language, you hold the power to select where the wave function collapses to create reality. The more often you rest in this gap, the freer you become. [For the path to Choiceless Awareness, see *Connecting with Your Soul* in Reflection 7.]

Principle 7: The World Is Your Mirror

Every relationship reflects a trait within you that is either loved or disavowed by you, and likely others as well. Attraction signals qualities that you happily and easily express. Aversion points to shadow material asking for change. When external circumstances displease you, ask yourself, *What inner adjustment would create a kinder reflection of myself?*

Principle 8: Evil and the Collective Shadow

"Evil" is the aggregate of unexamined dark shadows in individuals and societies. Holocausts and hate crimes begin as denied impulses and complicity. Healing starts at home within you: So acknowledge your own shadow, speak it to a trustworthy ear, and bathe it in your forgiveness. Because consciousness is nonlocal—Collective Consciousness—each private, positive act of integration lightens the load of the collective shadow.

Principle 9: Multidimensional Consciousness

We live in a universe of *multidimensional consciousness*, where our state of awareness determines our perception of reality.

Walt Whitman described this as a *Glimpse of the Soul*, writing: "I must not be awake for everything looks to me as it never did before, or else I am awake for the first time, and everything before was just a mean sleep."[45]

"I am awake for the first time" points to episodic openings into higher-energy states of consciousness. Beyond an ordinary state of awake, lie:

- **Cosmic Consciousness:** Your sense of self expands to include the Universal Self.
- **Unity Consciousness:** The observer and observed become one.

Quantum theory describes particles as both local and nonlocal; mystics report the same of awareness. Stillness and deep meditation provide a liminal entry, a threshold.

Principle 10: Death as Evolution's Engine to Life

Death is not an endpoint. It is the cosmic mechanism by which the universe evolves and elevates consciousness to higher energy levels. Your body is undergoing a process of death and renewal at this very moment. Your cells die by design so your organism may renew. Stomach cells regenerate every five days, skin cells every month—each reincarnating with the precise memory of their function.

Our bodies are incarnations of memory, imagination, desire, death, and quantum creativity. Even now, the universe is dissolving and reforming at the speed of light. Each cycle is an opportunity for evolutionary refinement—an upward energy trajectory of Universal Consciousness.

This *dual evolution*—physicality and consciousness—is the primary engine driving the universe's progression.[46] In biology, programmed cellular death is called *apoptosis*—without it, uncontrolled cell growth leads to cancer. Just as cellular death is necessary for physical health, death itself is essential for the continuous refinement of physical existence and higher levels of Universal Consciousness.

On a cosmic scale, stars explode to forge new elements. Physical death serves the same upward spiral, releasing consciousness to subtler higher energy planes. Recognizing this recasts death from an enemy to an essential ally.

Principle 11: The Power of Present-Moment Awareness

If we conquered death, we would all become *living mummies* in a fossilized universe, devoid of growth. *Quiet your mind and embrace the present.* Let go of the past—it is no longer here. Let go of the future—even though you may imagine it.

As Eckhart Tolle demonstrates in his book *The Power of Now*, learning to focus on the present dissolves fear and induces immense clarity, peace, and creativity. He concludes: "You are here to enable the Divine Purpose of the universe to unfold [see Reflection 9, Conjecturisms 5 and 8]. That is how important you are."

The *Principle of Inextricability*[47] states that physical evolution and consciousness evolution are interconnected and eternally interdependent. While we must learn from the past and plan for the future, our greatest power lies in anchoring ourselves in the present moment—through *Present-Moment Awareness.*

Principle 12: The Universe Sees Itself Through You

The calcium atoms in your bones and the iron atoms in your hemoglobin were birthed within the core of super-dense stars and dispersed throughout the universe in ancient supernovae. Your eyes are star-stuff fashioned into lenses so that the cosmos can witness its own artistry. A child's wonder echoes this primordial memory. My young daughter, once looking at the moon through a telescope, asked me,

"Daddy, do you think the moon can feel us looking at it?" A simple, amusing, yet profound comment.

Principle 13: You Are Not a Fixed Person

As the Buddha taught, "There is no such thing as a person." Both the observer and the observed are expressions of a deeper consciousness. When you recognize this, you gain the power to manifest the person you choose to be and the reality you seek to create. The key to this enlightenment is meditation, Choiceless Awareness, and Present-Moment Awareness.

Principle 14: Synchronicity: Messages from the Universe

Coincidences are not accidents—they are messages from the universe (Cosmic Consciousness). Each synchronicity carries significance. When one occurs, pause and reflect: *What is the universe communicating?* In stillness, the answer will emerge.

Principle 15: Seek Answers from Within

Keep the fundamental questions stated in Principle 1 alive, and focus on them daily: *Who am I? Why am I here? Where am I going?* And meditate daily. The inquiry itself is a beacon that eventually brings you home.

Putting It All Together

Choose the principle that stirs your heart. Sit with it, practice it, and test it against experience. Over centuries, these methods have proven reliable: They cultivate well-being, amplify creativity, and align personal desires with the evolutionary current flow of the universe.

Focusing frequently and mindfully—perhaps in a state of Choiceless Awareness—on any one of these principles has been proven over the millennia to create well-being, happiness, and success in life.

REFLECTION 9

True Reality and the Universe in a Nutshell

10 Conjecturisms[48] *on Life and the Universe*[49]

"When you want something,
all the universe conspires to help you achieve it."

—*The Alchemist,* Paulo Coelho[50]

All that has been presented so far relies on information drawn from the intersection of quantum and Transcendent Physics with Eastern Wisdom. Before transitioning to Conscious Quantum Physics, it's useful to succinctly summarize, through the following 10 Conjecturisms, key technical conclusions based on Transcendent Physics.

Conjecturism 1: Birth of the Universe

Our universe originated from an event of significant magnitude—the "Big Bang"—emerging from an unfathomable realm of infinite potential into four-dimensional space-time. From a singularity smaller than an atom and containing "something," an immense, explosive expansion unfolded with staggering speed, giving rise within the universe to *all* of the energy and matter we know of today.

Conjecturism 2: The Primacy of Consciousness

Beneath the physical structure of the universe lies its greatest mystery. *In a cosmic process known as Universal Transubstantiation,*[51] *that fundamental "something" that emerged from the singularity was not a mere physicality, but* consciousness *itself—the very essence and source of all existence and the creative force behind the formation of all of the energy and mass in the universe and its continuous evolution.*[52] The laws governing consciousness transcend classical and quantum physics and are increasingly explored in the emerging field of Transcendent Physics.

Of all current models for the science of consciousness, the closest to that embraced by Transcendent Physics, is Panpsychism:[53,54] *Consciousness is intrinsic and ubiquitous throughout the universe, and all energy and mass contain some level of consciousness.* However, a refined version of Panpsychism, which more accurately supports the science of Transcendent Physics, is called *Trans-Immanent Panpsychism:*[55] *Consciousness is intrinsic and ubiquitous throughout the universe, and also exists beyond all space-time.*[56]

Some identify the highest level of consciousness, Cosmic Consciousness, as a Divine Source, or God. Some philosophers and many theologians argue that Its presence pervades all things, uniting the cosmos in a silent, synchronous, and eternal rhythmic force.[57] For example, although related, there is a subtle difference between *pantheism ("All is God")* and *panentheism ("All is in God, and God is in all things").*[58,59]

Conjecturism 3: The Spectrum of Consciousness

Consciousness is not merely a function of existence—it is True Existence, itself. As Wisdom Seekers of the East say, "It is not just an important thing—it is the only thing!" Consciousness reveals itself throughout the universe in a triad of expressions:

- **Immanent Consciousness:** The foundational awareness inherent in every energy field and particle of matter throughout the universe. It is infinite and eternal—the True Reality of all living and nonliving species.

- **Collective Consciousness:** An interconnected web of awareness linking all beings and entities, forming a universal network of unity throughout the cosmos.
- **Cosmic Consciousness:** An infinite and eternal Source, embodying Six Cosmic Core Values: truth, wisdom, understanding, forgiveness, love, and compassion. These core values are not, in any way, a *moral injunction*. They are fundamentally necessary for the universe to effectively and efficiently achieve and express its Meaning and Purpose (see Conjecturism 7). Omniscient, omnipotent, and omnipresent, Cosmic Consciousness guides the continuous unfolding of Universal Meaning and Purpose via its presence in all energy and matter throughout the universe.

Conjecturism 4: Our Immanent, Transcendent, Nondualistic Universe

We inhabit a nondualistic universe in which Immanent, Collective, and Cosmic Consciousness permeate everything and reside within all species and entities. This interconnectedness allows for amazing possibilities—individually, collectively, and cosmically.

To access this inner potential, one must quiet the distractions of fear, desire, and worldly chaos, preferably through deep meditation or other consciousness-elevating practices. In contrast, *dualistic* perspectives perceive separateness rather than unity, which is the root of humanity's most perilous challenges.

Two important states of being prevail: *Transcendence* is a state of being *beyond* our physical world, while *Immanence* refers to a state of being *within* our physical world. To say the Divine is Transcendent means that the Divine is beyond us in space and time. To say that the Divine is Immanent means that although the Divine is present in space and time, Its presence is also *within* us.

Most organized religions emphasize the Transcendence of the Divine—that is, duality: "We are down here, and the Divine is up there."

The Consciousness Conjecturisms embrace the Transcendence state of being but strongly emphasize Immanence (nonduality): "The Divine is here within us."[60] Therefore, the Divine is at once both Transcendent and Immanent—providing us with a cosmic source of great possibilities.[61]

Since our universe functions in accordance with the principles of nonduality, at its core, all energy and matter in the universe are interconnected, although the intensity of this connection may vary, based on the energy levels of the participatory elements of consciousness. Other determinants are intention, attention, distance, time, and personal affinity.

Because of this connection, through alignment with the Core Values of the universe, and by accessing elevated altered states of consciousness via deep meditation, it is readily possible to use this intimate communication for personal manifestation, provided that it adheres to the Laws of Manifestation. Unfortunately, the unique power of manifestation is not recognized and accessed by very many people.

Conjecturism 5: Universe Creation, Dissolution, and Re-creation

Only about 5 percent of our universe is visible; 26 percent consists of invisible *Dark Matter*, which organizes and stabilizes the cosmic structure of the galaxies of the universe, while 69 percent is invisible *Dark Energy*, also known as the *Cosmological Constant*, and is driving the universe's accelerating expansion.

Conjecturisms 1, 2, and 3 imply a fundamental realization: *In the beginning, there was no beginning.* Transcendent physics posits that in the next several billion years, Dark Matter and, particularly, Dark Energy, will weaken sufficiently for their net gravitational forces to halt the universe's expansion, ultimately reversing it into contraction. This scenario culminates in what some physicists call the *Big Crunch*[62]—all of the mass and energy converting back to consciousness, creating a singularity akin to that from which the universe emerged.[63]

From this new singularity, another cosmic expansion will initiate creation of the next universe, an overall process known as the *Big Bounce.*

This cyclic process repeats eternally, with each successive universe preserving and advancing the information and consciousness of previous ones. Over time, the universe evolves, ascending to higher levels of complexity and awareness.

Conjecturism 6: Human Uniqueness

Among all of Earth's species and entities, humans stand at the pinnacle of both physical and consciousness evolution, although all species and entities contribute to the grand evolutionary journey.

Conjecturism 7: The Meaning and Purpose of the Universe

Guided by the Six Cosmic Core Values, the universe is engaged in a continuous, synchronous evolutionary dance of physicality and consciousness. These twin processes are inextricably intertwined, inseparable, and mutually interdependent—the universe would not exist without their mutual interplay.

The *Meaning* of the universe is its ongoing transformation, while its *Purpose* is the continuous elevation of its consciousness energy. Consciousness is inherently perfect. The only variable metric is its energy level, which can vary with distance, time, and the physical and spiritual state of an entity.

As with the *Universal Laws of Conservation of Energy and Mass*, consciousness, too, can be neither created nor destroyed—only transformed into energy or matter.

Conjecturism 8: The Meaning and Purpose of Life

The Meaning of Life for each being—irrespective of species—is to recognize and embrace its inherent gifts. Its Purpose is to apply those attributes, guided by the Six Cosmic Core Values, in service of the universe's physical and consciousness evolution to higher energy levels, thereby fostering a more harmonious, sustainable, and elevated existence for all.

Conjecturism 9: The Reality of Reincarnation[64]

To sustain the eternal cycle of cosmic formation and rebirth, and given the infinite and eternal nature of consciousness, Transcendent Physics aligns with Eastern Wisdom in positing *reincarnation* and its associated *karma* as a fundamental mechanism of universal renewal. Karma is not a reward or punishment consequence for good deeds or misdeeds, respectively, but a natural result of current and prior actions guided by the Six Cosmic Core Values, thereby creating a consistent cosmic energy balance, which is congruent with the Cosmic Core Values.

Conjecturism 10: Laws of Manifestation

Any manifestation should align with the universe's purpose and should adhere to these two principles:

- **Nonviolence:** It will cause no harm by adhering to relevant ethical norms governing human and nonhuman life.
- **The Greater Good:** It will further the world, however modestly, toward improved conditions reflecting the Six Cosmic Core Values.

Through this means of manifestation, you cannot help but succeed, because you have placed your sanctioned intentions into the vast universe of all possibilities and enabled Cosmic Intelligence to work through you for the most effective outcomes.

These 10 Conjecturisms are intended to inspire curiosity, invite deeper exploration, and illuminate our understanding of ourselves, our world, the universe, and the vast consciousness that binds us all together.

REFLECTION 10

What Is Consciousness, Really?

"I think death is just a transition to another state of consciousness."[65]

—Val Kilmer, actor

In Part II, we explore the outcomes of Conscious Quantum Mechanics. Our further analysis concerning the functioning of consciousness is central to that exploration. Although some may disagree with our view, it's useful to provide a perspective for consciousness as it specifically applies to Transcendent Physics. There are numerous models for consciousness. We will focus on the one that supports the principles of Transcendent Physics. Excellent treatments for many of the most important models can be found in texts by Annaka Harris[66] and Philip Goff.[67,68]

This reflection explores consciousness through the lens of Trans-Immanent Panpsychism, a spinoff of Panpsychism and a model that posits that consciousness is intrinsic to all energy fields and matter in the physical universe and also that it exists beyond space-time.[69] Within this framework, consciousness operates not just as an awareness or experience but as the *intrinsic* foundational substance of the cosmos, pervading every aspect of reality.

Trans-Immanent Panpsychism: A New Paradigm

Trans-Immanent Panpsychism bridges the large gap between all tangible and intangible things, connecting the fundamentally scientific with the profoundly philosophical, by proposing that consciousness is an intrinsic quality of the universe, extending beyond space-time and pervading every energy field and particle of matter created during the Big Bang—and, in fact, it submits that consciousness is the source of all energy and matter.

Traditional sciences often question how consciousness could suddenly arise from insensate matter. Trans-Immanent Panpsychism circumvents this conundrum by suggesting that all matter and energy possess some degree of consciousness, a legacy from the initial subatomic singularity, the point of pure consciousness that catalyzed the birth of our universe.

This hypothesis challenges the notion that consciousness suddenly emerges at some point within higher levels of species complexity. Instead, it suggests that consciousness exists in rudimentary forms within the most basic embodiments of energy and matter. This view directly challenges the traditional strong *emergence* theories in science, which have struggled unsuccessfully to explain consciousness purely through the integration of physical systems with the neurological sciences.

Famed biologist and polymath J. B. S. Haldane had this to say about emergence theories: "If consciousness were not present in matter, this would imply a theory of strong emergence that is fundamentally anti-scientific. Such emergence is radically opposed to the spirit of science, which has always attempted to explain the complex in terms of the simple. . . . If the scientific point of view is correct, we shall ultimately find them (signs of consciousness in inert matter), at least in rudimentary form, all throughout the universe.[70]

A New Perspective on Consciousness

Trans-Immanent Panpsychism offers a fresh viewpoint that directly links the tangible and intangible aspects of our universe. It suggests that consciousness is an inherent characteristic of the universe itself, extending beyond the fabric of space-time and existing in all forms of energy and matter that originated from the Big Bang. According to this theory, consciousness is not just a product of complex biological structures, but rather a fundamental essence present in everything.

This perspective challenges the belief that consciousness could only arise in more complex beings. Instead, it asserts that even the simplest particles are imbued with a basic level of awareness. This stance stands in opposition to mainstream theories suggesting that consciousness somehow emerges only at higher levels of complexity, yet none of these theories have yet to clarify when, why, and how this transition occurs.

Contrasting Theories: Complexity Emergence vs. Trans-Immanent Panpsychism

Understanding consciousness can be framed through two distinct lenses:

1. **Complexity Emergence:** This viewpoint claims that consciousness arises only when matter reaches a certain level of complexity, such as in the human brain. It suggests that prior to this threshold, there is no awareness or consciousness present.

2. **Trans-Immanent Panpsychism:** In contrast, this theory posits that even the simplest constituents of matter possess some form of consciousness. While this does not imply that atoms have thoughts, it means that they carry a form of rudimentary "awareness." As matter evolves into more complex systems, this inherent consciousness becomes more organized and powerful.

Consciousness does not appear abruptly; rather, it has always been present, intrinsic, and embedded within the cosmos. As simple ingredients combine and form more complex structures, their inherent consciousness amplifies, synergizes, and transforms—creating new, richer experiences that exceed mere simple addition.

This concept is mirrored in numerous biological systems where the collaboration of parts produces outcomes unattainable by individual parts alone.

Examples of Emergent Systems

1. **Neurons to Mind**

- **Basic Unit:** A single neuron transmits signals on its own but cannot think.
- **Complex Network:** When billions of neurons link together to create the human brain, they form intricate networks. This connectivity facilitates cognitive functions such as thought, emotion, and memory—a human mind.
- **Emergent Properties:** The mind emerges not merely as a by-product of neurons but as an intricate expression of the deeper consciousness present in all matter. This individual consciousness connects to a broader collective consciousness that permeates the entire universe.

2. **Amino Acids to Life**

- **Fundamental Elements:** Amino acids are the simple basic chemical building blocks of all proteins.
- **Synergy Formation:** Linked together, amino acids form long polypeptide chains that fold into highly complex structural proteins, which are essential for life.
- **Emergent Functions:** These proteinic macromolecules provide critical roles—acting as *enzyme catalysts* (for example, DNA helicase). They exponentially accelerate necessary biochemical reactions. As *structural components*, they provide bodily stability and longevity (for example, collagen); as *transporters*, they deliver necessary biomolecules throughout the body (for example, hemoglobin); and as *messengers*, they send, in a timely manner, important biomolecules (for example, insulin) required for excellent health and peak performance. A case in point: Enzyme catalysts speed up reactions by millions to billions of times. Without these molecular machines, life as we know it would be virtually

impossible. Other examples can be cited for structural components, transporters, and messengers. The transition from simple amino acids to essential life functions illustrates how complexity not only organizes but also amplifies the outcomes of underlying consciousness.

Conclusion: The Universe as an Expansive Consciousness

While theories of complexity emergence view consciousness as a fortunate by-product of advanced brain functions, Trans-Immanent Panpsychism considers it a fundamental quality of existence. Consciousness is seen as evolving alongside matter, growing in sophistication as systems become more complex.

In this view, every component of the universe, from subatomic particles to vast galaxies, contains some form of consciousness. The distinction between insects, animals, human beings, and what we might call the Divine or Cosmic Consciousness lies not in different types of consciousness, but in varying energy levels of the same universal life force. The essence of consciousness is unified throughout all existence.

The Consciousness Conundrum and the Combination Problem

A central issue in all Panpsychism models is the *Combination Problem*: If the universe's smallest constituents possess consciousness, how does complex consciousness emerge in intricate forms, like human experience? As mentioned above, *Trans-Immanent Panpsychism contends that consciousness does not merely linearly sum up; it exponentially synergizes during the combination process.*

That is, consciousness in higher entities does not equal the mere sum of simpler low-energy consciousness units but rather results from intricate interactions and combinations that transcend simple summation. Unlike the energy fields of classical and quantum physics, consciousness energies don't aggregate as a simple sum of the parts—*the whole is well beyond the sum of the parts. The extent beyond is a function of the level of consciousness in the participating parts, the available synergies by combining these specific parts, and the degree of summation.*

Although the combinatorial rules for these enhanced synergies have yet to be quantitatively determined, there are numerous examples where such synergies have been observed in complex systems—for example, in ecosystems, weather patterns, economies, social networks, and the human body. Perhaps a useful metaphor is: If you take apart a watch, none of the parts can tell time, yet when put together in a specific way, they yield an accurate timepiece. Drawing on this metaphor, the yet-to-be-determined combinatorial rules will likely hinge on *how* the combination occurs—that is, the order in which the parts combine.

Transcendent Physics posits that *summing consciousness of various energy levels does not change the amount of consciousness, only its final energy level. The total amount of undifferentiated consciousness in the universe plus that within its surrogates (matter and energy) is constant.*

Most neuroscientists regard consciousness as emerging from complex processes within the brain. While neuroscience has made significant strides in correlating neural activities with behavioral phenomena (neural correlates), it remains unable to explain the qualitative experience, or qualia, that define human consciousness.

This is the classic Hard Problem of consciousness—the enigma of subjective experience—how do neurological processes account for the experiential versus sensual occurrences of anything, such as sight, smell, taste, touch, joy, sorrow, awe, or a thought? What's a thought, where does it come from, and where does it ultimately go? What are the synaptic translations of feelings—for example, feeling that something is breathtaking, or the colorful spectrum of sensory perception—that is, seeing green?

The Hard Problem can be broken down into two parts: *sensual* and *experiential* occurrences. For example, let's say you've traveled to London for a vacation and are staying in a beautiful luxury hotel. You wake up in the morning, excited to start your day, and looking out your window you see fog and a soft falling rain.

Assuming you have reasonably good eyesight, you will likely detect this *sensory* occurrence—fog and rain—a message that neuroscience can explain in great detail based on your ocular sensory system and numerous neurological synapses on the way from your eyes to your brain, a complex sensory process.

However, it cannot provide a neurological mechanism for your subjective *experiential* feelings, which, for example, might be, "How lovely! Now this is the London I remember!" Or, alternatively, it might be, "Damn! I wanted to walk through Hyde Park today and have a picnic lunch there. Now what am I going to do?" These are obviously two very different *experiential* occurrences, which neuroscience has yet to come even close to explaining how, where, and why they occur.

To date, neuroscience has been unable to elucidate these facets, where the overt physical realm converges with the *intangible mental experience*. This impasse suggests that traditional sciences may falter without embracing metaphysical perspectives akin to those embraced by Transcendent Physics.

Revisiting Science: Navigating the Metaphysical Divide

The Scientific Method, with its deep roots in empirical observation and rigorous replicability, faces a pronounced barrier in addressing the metaphysical nature of consciousness. Yet, the acceptance by many scientists of seemingly far-fetched, multiverse, and multidimensional theories—such as the Many Worlds Theory and various String Theories—highlights the paradigm whereby metaphysical intuition interlaces with scientific exploration without empirical substantiation, and with little promise that such substantiation lies anywhere on the research horizon. And yet, more than a few quantum physicists thrive on these theories.

Progressing along this premise, Trans-Immanent Panpsychism suggests that consciousness, as an intrinsic entity, not only underlies the universe but guides its evolution—physically and consciously. The enigmatic essence of veracious intuitions or senses of correctness without rationale evokes an underlying consciousness, subtle yet ubiquitous. It is the permeating current guiding interactions at both the microcosmic and macrocosmic levels, intertwining physical actions with purposeful consciousness pursuits.

Moreover, consciousness as the root of veracious intuition—a sense that something is correct without a known rational explanation—demands an understanding that transcends the conventional boundaries of science. It opens avenues for exploring how consciousness informs decision-making, moral behavior, intuition, and creativity, urging a deeper investigation into its potential as more than a mere by-product of neural activities but as a powerful tool to explain existence. These forms of intuition occur almost exclusively at higher levels of consciousness such as those accessed as altered states during deep meditation practices. In essence, Trans-Immanent Panpsychism posits that deep meditation provides a means to access high energy levels of consciousness, thereby enabling one to tap into the Infinite Wisdom and Knowledge of the Universe.

Consciousness, Free Will, and the Human Experience: The Pinnacle of Evolution

While consciousness exists at varying levels of energy across all entities, on planet Earth humans uniquely represent the pinnacle of consciousness evolution, capable of accessing profound depths of awareness. The human brain-mind connection acts as a mediator, a "control valve," regulating consciousness flow in alignment with dimensional expressions of reality and inner personal sensitivity. This mediating mechanism allows humans to traverse beyond immediate experience, expanding their understanding—sometimes even beyond space-time—through altered states, induced, for example, by meditation—a practice that harnesses these same pathways to align with higher cosmic energies generated in coherence with the Six Cosmic Core Values.

Consciousness fuels human potential, steering choices within the realm of free will. Yes, Transcendent Physics and Trans-Immanent Panpsychism embrace and encompass the concept of free will, however, recognizing that it can be moderated, sometimes significantly, by external forces. Unveiled consciousness—a direct line untainted by the world's noise, chaos, and distractions—enables clarity of purpose and action, aligning endeavors *to continuously increase the total energy level of consciousness throughout the universe—an infinite and eternal process and, according to the science of Transcendent Physics, the very purpose of the universe.*

Humans have access to pure consciousness, which is the source of free will. However, neurological changes within the brain, whether purposeful or inadvertent, can inauspiciously alter not only the access to higher levels of consciousness but its translation as well. At low energy levels of consciousness, the brain, in conjunction with its history and prior programming, as well as world events, can alter free will. At increased energy levels of consciousness, this impact can be minimized or completely eliminated, leading to the expression of pure, unmodified free will.

Expanding the Notion of Consciousness Through Case Studies

Despite the prevailing resistance to metaphysical approaches, numerous concrete, vetted expressions of consciousness, such as near-death experiences (NDEs), out-of-body experiences (OBEs), and spontaneous healing, present compelling evidence for its transcendent nature. These phenomena depict consciousness operating beyond measurable physical causes, reinforcing Trans-Immanent Panpsychism's postulate that consciousness is an all-powerful force, essential to a comprehensive understanding of reality.

- **Near-Death and Out-of-Body Experiences:** NDEs[71,72] and OBEs[73] provide intriguing glimpses into the realm of consciousness that transcends physical boundaries. Many individuals who have undergone an NDE or an OBE report profound experiences of awareness and clarity outside their physical bodies. These accounts often include encounters with a pervasive "loving presence" (Cosmic Consciousness); memories of detailed life reviews; vivid, detailed, mentally retrievable awareness during long-distance travel (the basis for psychic espionage[74]); and journeys through radiantly impressive landscapes.

  Such narratives challenge purely biologically based neurological explanations of consciousness, suggesting a continuation of awareness beyond bodily functions and existence. Increasing documentation and analysis of NDEs and OBEs push the limits of—and likely transcend—the capabilities of conventional neuroscience, inviting a broader, more open approach in understanding consciousness.

- **Spontaneous Healing:** Another area where consciousness may extend beyond its physical confines is in instances of spontaneous healing.[75,76,77] Medical literature abounds with numerous cases where individuals recover from severe illnesses without medical intervention and clear explanations. Often, these healings follow intense personal experiences with significant shifts in consciousness, such as deep meditation, visualization, or fervent prayer. These instances suggest that consciousness can potentially influence physical health in significant ways, highlighting an area rife for further scientific inquiry and acceptance.

The Hierarchies of Consciousness

In Transcendent Physics, consciousness is expressed in three key forms: Immanent Consciousness, Collective Consciousness, and Cosmic Consciousness. This delineation is not meant to indicate a hierarchical superiority among the three forms, but rather different dimensions of function, levels of energy, and scales of consciousness at work throughout the universe.

- **Immanent Consciousness:** This is the fundamental awareness that pervades every iota of matter and energy. It is the consciousness inherent in the universe's building blocks, embedding each atom and molecule with awareness. This consciousness is infinite, nonlocal, and eternal, providing a baseline of awareness that defines True Reality.
- **Collective Consciousness:** It emerges from the interaction and interconnectedness of individual Immanent Consciousnesses. It refers to the shared awareness that binds creatures and entities, forming a contiguous web of consciousness that spans the entire universe. In principle and practice, this interconnectedness enables unity—fostering shared experiences, societal growth, and cultural evolution.

- **Cosmic Consciousness:** This is the most expansive expression because it has the highest energy level of consciousness, reflecting an infinite, eternal Divine Source. This consciousness resonates with the Six Cosmic Core Values. These values are not arbitrary moral injunctions; they are fundamental to, and necessary for, the universe's purposeful development to higher energy levels, marking the navigation points for both individual and universal evolution.

The SOP Conjecturism: Beyond Reductionism

The Sum of the Parts (SOP) Conjecturism suggests that the consciousness level of a complex entity is greater than the mere aggregation of its components' consciousness. For instance, the sum of consciousness within the 33 trillion cells forming a human being does not equate directly with human consciousness. Instead, the consciousness of complex entities arises from synergies, interactions, and unique configurations beyond that of a simple sum of the parts' contributions.

A simple molecular metaphor might help. The molecule limonene exists in two prominent forms, each a mirror image of the other: d-limonene and l-limonene. Both contain the same number of identical atoms; however, they are arranged in a slightly different spatial configuration, and each has properties different from the other. D-limonene is found in citrus fruits, giving them their distinct characteristic scent and flavor of lemons and oranges. However, l-limonene has the scent and flavor of pine.

Key Transcendent Physics proposals behind the SOP Conjecturism include:

- **Cumulative Karmic Influence:** Each individual harbors consciousness shaped by past lives and karma, subordinate to universal laws of energy and mass conservation. This karmic influence plays a pivotal role in defining consciousness energy levels. It is neither a reward for good deeds nor a punishment for misdeeds. It's a parallel to Newton's Third Law: *Every action, whether tangible or intangible—has an* *equal and opposite reaction.* Karma is a mechanism for maintaining the cosmic balance necessary for Universal Purpose.

- **Evolutionary Complexity:** Genetic and epigenetic factors influence consciousness, but their interaction with karma and cosmic principles adds layers to consciousness well beyond genetic coding.
- **Cyclical Universal Evolution:** Universe creation occurs in a cyclic process via expansion, contraction, dissolution, the formation of a unique consciousness-containing singularity, and its subsequent expansion to create the next universe—each time cumulating and preserving total information and consciousness over successive iterations. Each universe in this infinite cycle carries forward a legacy of consciousness and information from its predecessors, contributing to the overall ascent and complexity of universal consciousness.

Philosophical Insights: Transcendence and Immanence in Consciousness

Exploring consciousness through the science of Transcendent Physics invites a reflection on the concepts of Immanence and Transcendence, each of which reference either a dualistic or nondualistic universe. Transcendence is a state of being beyond our physical world, whereas Immanence is a state of being within our physical world. To say that the Divine is Transcendent means that It is beyond us in space and time. To say that the Divine is Immanent means that It is both present in space and time and an intimate part of us. These states of being are particularly important when comparing Transcendent Physics with organized religion.

Laws of Manifestation: Consciousness as a Creative Force

As discussed earlier, any manifestation aligning with the universe's purpose must also abide by two fundamental principles:

- **Nonviolence:** It must cause no harm, adhering to relevant ethical norms governing human and nonhuman life.

- **The Greater Good:** It should further the world, however modestly, toward improved conditions.

Conscious intention, attention, and elevated consciousness states can facilitate meaningful manifestations, aligning personal aspirations with the universe's evolution, purpose, and meaning.

Conclusion: Embarking on a Conscious Journey

The universe is intricately engaged in elevating its universal consciousness energy through ongoing interactions and transformations. In embracing Transcendent Physics and its broader paradigm of consciousness, we are asked to step beyond the confines of traditional scientific thought and embrace a more integrative approach.

This journey involves recognizing consciousness not merely as a complex outcome of neural processes but as an essential component of reality, acting both locally within entities and nonlocally with a universal guiding force. This means that certain aspect of metaphysics are necessary for the metamorphosis of conscious quantum physics to the field of Transcendent Physics—echoing the historical transition of classical to quantum physics—to pave the way for the growth of this field.

The Experiential and Experimental Exploration of Consciousness

As we explore consciousness scientifically and experientially, several pathways emerge for deeper exploration:

1. Scientific Inquiry

- **Interdisciplinary Research:** Bridging neuroscience, physics, and philosophy to understand consciousness requires interdisciplinary approaches. Initiatives integrating the fields of cognitive science, neurobiophysics, quantum physics, and metaphysics could unlock new insights into the consciousness phenomenon.

- **Emergent Technologies:** Leveraging emerging technologies, such as AI and machine learning, could unearth patterns and correlations in consciousness-related phenomena, promoting a broader understanding of nonphysical dimensions, and suggesting further experimentation.

2. Experiential Practices

- **Meditation and Mindfulness:** Practices that foster deep self-awareness and transcendental states could enhance personal understanding of consciousness. Techniques like meditation and Choiceless Awareness provide insights into the multilayered and interconnected aspects of individual and collective consciousness. Similar to studies that have been carried out at Harvard[78] and MIT,[79] it is valuable to access the knowledge and experience of long-standing meditators, particularly those who have practiced protracted meditation—for example, Buddhist monks.

- **Induced Meditation:** It may be possible to induce within the human brain an electromagnetic field of precise frequencies that instantaneously shuts down the "monkey chatter" of the normal brain to create a "quiet" (the "gap"), which often only occurs after many years of deep meditation practice. Hypnosis may be a helpful complementary adjunct. However, this must necessarily be pursued with great caution to assure that the end result is perfectly reversible to the normal state of the brain. This technology would provide more expedient and broader access to high levels of consciousness.

Consideration should be given to prudent exploration of chemically induced meditative states via careful administration and study of mind-shifting agents such as LSD (lysergic acid diethylamide), psilocybin/psilocin (4-hydroxy-N,N-dimethyltryptamine), mescaline (3,4,5-trimethoxyphenethylamine), ibogaine (12,13-Dimethoxyibogamine), ketamine[2-(2-chlorophenyl)-2-(methylamino)-cyclohexanone], MDMA/Ecstasy (3,4-methylenedioxy-methamphetamine), DMT (N,N-dimethyltryptamine), and 5-MeO-DMT (5-methoxy N,N-dimethyltryptamine). Most of these chemical agents, while of potential value for controlled and effective mind-altering outcomes, can have clinically perilous, even lethal, results; and therefore, beyond legal consent and oversight, will require research that is managed by knowledgeable and experienced scientists.

Michelle Carr is an assistant professor of psychiatry at the University of Montreal. As director of the Dream Engineering Lab at the Center for Advanced Research in Sleep Medicine, she and her team have demonstrated, via induced lucid dream technology, a relatively straightforward means to achieve altered states of consciousness as a potential link to "mystical experiences on demand."[80] This technology may demonstrate safe, ready access to cooperative functions between Immanent and Cosmic Consciousness.

- **Neuromodulation Techniques:** These explore how various modalities—for example, neurofeedback, transcranial magnetic stimulation, biofeedback, functional MRI, magnetoencephalography, electroencephalogram, and vagus nerve stimulation—alter or correlate with consciousness levels, offering practical applications for enhancing cognitive, emotional, and psychic well-being.

Potential Impacts on Society and Science

Embracing a consciousness-centered framework might transform various social domains, as outlined below:

- **Health Care and Wellness:** Recognizing consciousness' role in health could revolutionize medical practices, promoting holistic treatments that incorporate mental, emotional, and spiritual dimensions that complement physical health interventions.[81]

- **Education and Development:** A consciousness-grounded education system that is initiated early in the curriculum—for example, elementary school—could nurture individual potential, empowering students with a deeper understanding of themselves and their interconnected roles within society. This would likely require a creative technique to teaching and therefore a new approach to educating teachers.[82]
- **Environmental Conservation:** By acknowledging consciousness' pervasive presence in nature, society might foster more profound ecological stewardship, approaching planetary crises with a harmonized focus on sustainability and interconnected survival.[83]
- **Philosophical and Personal Reflections:** The implications of understanding consciousness through Trans-Immanent Panpsychism invoke profound philosophical and personal reflections. *Humanity's place within the cosmic order shifts from being isolated observers to participants actively and collectively shaping universal evolution.* Each action taken is karmic in its essence and resonates with cosmic meaning, encouraging personal and communal journeys aligned with core universal values.[84]
- **Ethical Considerations:** The awareness of consciousness' presence in every particle advocates for an ethics of respect and care not just for sentient beings but for the broader environment. Actions take on greater significance when framed against this backdrop of interconnectedness and responsibility to the collective consciousness of the whole.[85]
- **Personal Growth:** Embracing a consciousness-centered view empowers individuals in their personal and spiritual growth journeys. Realizing interconnection and embeddedness in universal consciousness inspires practices fostering inner peace, understanding, compassion, and purpose-driven living. Inner changes are almost always reflected externally.[86]

The Future of Consciousness Exploration

The trajectory of consciousness exploration demands courage to traverse and connect uncharted scientific and philosophical territories. The principles of Transcendent Physics expand traditional frameworks, offering a sophisticated and integrative approach to scientific inquiry and existential understanding.

Challenges and Opportunities

- **Overcoming Skepticism:** The greatest barriers remain philosophical and institutional conservatisms preventing acceptance of nonmaterialistic explanations. Alleviating these through dialogue and demonstrated inquiry success might open the doors for mainstream acceptance. Demonstrating even minor success might go a long way toward diminishing skepticism.

- **Integrative Approaches:** Continuing to blend empirical research with experiential insights aligns science with transcendental experiences, enhancing empathy, connection, and understanding.

Conclusion: Consciousness—a Universal Dance

Ultimately, consciousness is an unending cosmic dance—a perpetual interplay of self and cosmos, inner and outer worlds, evolution and transformation. Transcendent Physics doesn't just ask us to understand this dance; it invites us to participate actively, harmonizing our movements with the rhythms of Cosmic Consciousness.

The insights gained from exploring the rich tapestry of consciousness will continue to shape not only the sciences and humanities but the very way in which humans relate to themselves, to each other, and to the universe. As we deepen our exploration, consciousness remains the vast, open field where science meets spirituality, inviting us to unravel mysteries that are as old as time itself.

The invitation stands: to engage with consciousness as both a scientific frontier and a philosophical quest, understanding it as the ultimate bridge connecting all facets of existence. Through this lens, the universe is seen not merely as a collection of separate parts, but a coherent whole—enriched by the silent interplay of all consciousness united in a cosmic symphony.

PART II

A Primer on *Conscious* Quantum Mechanics

Solvay Conference on Quantum Mechanics, 1927

"Observations not only disturb what is to be measured, they produce it."

—Pascal Jordan, developer of modern quantum theory

"I have thought a hundred times as much about the quantum problem as I have about general relativity theory."

—Albert Einstein

REFLECTION 11

Profound Perspectives

"I like to think the moon is there even
if I am not looking at it."[87]

—Albert Einstein

A Few Thoughts

It's not hyperbole to say that nearly all of the most advanced technologies in the world could not have happened without quantum mechanics. It can also be said that it's one of the most challenging sciences to understand, in part, due to its technical and mathematical complexities, but perhaps more so, because of its mystical and metaphysical implications. Most scientists *were* not, and for the most part, are *still* not, technically and philosophically trained to pursue this path without encountering stress and disappointment. Fortunately, some are willing to take the professional leap and personal risks to do so.

That said, I encourage you to make your way through Part II. I have tried my best to keep things simple, yet clear and concise, without sacrificing the accuracy of quantum physics. You will unquestionably come away with stronger context for the primary messages of this book and a deeper understanding of the nature of consciousness and its implications for your well-being and for humanity in general.

Although the emerging field of Transcendent Physics is a science that proposes to elucidates the nature and function of True Reality throughout the universe, just as classical physics was a key transition to quantum mechanics, the latter is an important stepping stone to Transcendent Physics—particularly in a format that can be considered to be Conscious Quantum Mechanics. Therefore, to grasp the truly profound implications of Transcendent Physics, it's important to understand its links to Conscious Quantum Mechanics.

Without any doubt, quantum mechanics has revolutionized our world since the fundamentals and applications of this science began to be uncovered around 1900—and continues to do so as we discover many of its unique mysteries. Whether you use a cell phone, surf the internet, have laser surgery, listen to a podcast, or watch TV, you are intimately dependent on the science of quantum mechanics, as is 40 percent of our global economy. This area of physics, especially as it relates to consciousness, is likely to revolutionize computing, AI, and cybersecurity in the form of quantum computing, quantum cryptography, and broad-spectrum artificial intelligence.

The Quantum Conundrum

What makes quantum mechanics so fascinating is that you can be highly trained and skilled in the mathematical and physical methodologies necessary to use the subject in a practical and constructive manner, yet if you confront the meaning and implications of quantum mechanics, it can seem impractical and disruptive in the context of what we perceive to be the "real" world. It brings physics face-to-face with metaphysics, and that's stressful to most scientists because it's incongruent with the Scientific Method, the sacred foundation for all good scientific research in the 21st century.

Think for a moment about how much funding a university physics professor might receive from the National Science Foundation for a research proposal titled "The Impact of Immanent and Collective Consciousness on Quantum Processes." Or consider how the audience at an American Physical Society meeting might react to questions about how consciousness might influence experimental results involving quantum systems. Or the odds that an assistant professor might have of getting promoted with tenure if their research embraced the interface of consciousness and quantum mechanics.

Before considering the implications of this conundrum—though it need not be a conundrum at all in view of its potential to significantly improve our quality of life—it will be helpful to review a few physical concepts that we will use along the way. For a preliminary understanding of basic quantum phenomena, we can consider Bohr's classical planetary model for the atom, discussed earlier.

Bohr's Quantum-Classical Model of the Atom

Picture the nucleus of an atom in a centric position like our sun. It is composed of noncharged neutrons and positively charged protons. The neutrons and protons are made up of subatomic particles known as quarks, leptons, and bosons. In Bohr's neutral atom, the positive charge of the protons is balanced by an equal number of negatively charged electrons spinning around the nucleus in much the same way the planets in our solar system revolve about the sun.

The electrons spinning around the nucleus can change orbits through interactions with a variety of forces, including photons—the basic particles of light, or more precisely, of electromagnetic radiation. For example, a low-energy electron closely circumnavigating the nucleus can be stimulated by a high-energy photon to jump up to a higher energy orbit. Or the reverse can occur: An electron can jump down to a lower energy orbit with a consequent emission of energy, usually in the form of a photon.

Using this model, often called the Old Quantum Mechanics, Bohr was able to derive a number of useful and practical conclusions. For example, he could calculate the line-spectrum for the hydrogen atom, and since it is the most abundant atom in the universe, these data could be used to analyze many cosmic phenomena, providing valuable information on the structure and function of the universe.

However, although Bohr's theory was a great success, it only worked for the hydrogen atom. Therefore, to really dig deeper into the physics and metaphysics of quantum theory, we must make the transitional leap from the old to the new quantum mechanics developed in the 1920s by Werner Heisenberg, Erwin Schrödinger, Paul Dirac, and other brilliant theoreticians. However, a point of clarity: Atoms, electrons, neutrons, and protons are not solid particles, as these models might imply. They are more precisely nebulous packets of *information* and *energy*.

That being said, quantum mechanics can explain numerous properties of what our five senses perceive to be the "stuff" of our universe. Thus, it can provide a more accurate perspective for your True Reality—which, hopefully, at this point, you know is not your physical body—and thereby give you specific tools that can significantly enhance your quality of life and well-being based on proven results and implications of quantum physics' experiments and models.

Conscious Quantum Mechanics

In nearly every statement that follows, no mathematics and only elementary concepts in physics will be used, which means that this is offered to you, the reader, as somewhat of an article of faith. If you were willing to learn the necessary advanced mathematical tools and become facile in pertinent areas of modern physics, you could move beyond articles of faith and have concrete evidence to judge the veracity of these statements for yourself. Perhaps it will provide some level of comfort if you take note that most physicists would accept the statements that follow, although many would be uncomfortable with some of their implications.

The following non-obvious—and some might say, unbelievable—phenomena follow directly from modern quantum theory and are key elements of *Conscious Quantum Mechanics*:

- **Observer Effect:** *What you perceive as reality is actually something you create via your observation of the universe.* When you consciously observe something under a given set of circumstances, events occur and/or objects appear. These events and/or objects are directly created by you, the observer. Before your observation, the events and/or objects are only states of energy vibrating in an infinite domain of possibilities.

  This includes trees, buildings, cars, and family members—essentially everything in the universe. It is your observation that causes the specific physical possibility to materialize. This effect has been shown experimentally for subatomic particles and atoms, and more recently for some larger molecules. The conscious creation from these experiments is the same effect we experience to be reality, based on our five-sense perceptions. In other words, you are constantly creating what you perceive.[88]

- **Infinite Possibilities:** *What actually exists out there is an infinite number of possibilities in the domain of potentia.* This infinite number of possibilities can be described by a complex mathematical expression called the *quantum wave equation* or *quantum wave function* (often denoted by ψ), where you can think of the wave function as being spread out in space-time, describing all possible outcomes.

  The probability of your perceiving a given outcome—a specific event or object under a given set of conditions—can be calculated from the square of the wave equation, ψ^2. When you create a certain event or object via your observation, mathematically speaking, you have caused the wave equation to collapse from all possibilities to a single result—what you observe is what you get.

  The late physicist and Nobel laureate Eugene Wigner speculated that observation and instantaneous collapse of the wave function happens at the very last stage of the observation process, probably when personal awareness occurs. He also speculated that human conscious awareness might actually reach out in some unexplained manner and change the physical state of a system. All we know is that somewhere on the scale between subatomic particles, atoms, molecules, and big objects such as cars, buildings, and human beings, a mysterious unexplained process occurs whereby observation collapses the wave function to a specific possibility and observation. And, voilà, there it is!

- **History Creation:** *Quantum mechanics predicts that your observation not only creates your present reality but also creates the past history appropriate to that reality—that is, you actually produce something backward in time.*

The late quantum cosmologist John Archibald Wheeler suggested a delayed-choice experiment to determine whether or not one's observation actually could produce something backward in time. For a while, this was simply conjecture and not experimentally feasible. However, with further developments in quantum technology, Wheeler's experiment has since been performed numerous times and confirms that the required backward-in-time history is created by the observer.[89]

Scientists still find this difficult to comprehend or believe, even though Einstein reminds us that time and space—that is, space-time—are only convenient parameters created by scientists to help understand the nature of the universe. They are not absolute entities.

- **Entanglement:** *Small particles such as electrons, atoms, and molecules experience a mysterious effect called quantum entanglement, wherein a change in one particle can instantaneously—faster than the speed of light—affect another particle at a significant distance away, even though there appears to be no known force field between the two particles.*

For example, it's possible to have two electrons proximate to each other and paired together, with one spinning to the left and the other spinning to the right. If now, these electrons are separated by hundreds of kilometers, and the spin on one of them is reversed, the other instantly reverses its spin as well—faster than the speed of light—as if it "knew" when and what to do![90]

Extrapolation of the results of all experiments carried out to date suggest that this same entanglement effect would occur even if these electrons were separated by billions of miles, located in different galaxies. The communication would still be instantaneous and immeasurably faster than the speed of light. How this communication occurs is still a mystery to physicists. It is a peculiar phenomenon, which Albert Einstein called "spooky action at a distance." It threatens the very essence of his relativity theory, wherein information based on the electromagnetic spectrum—for example, photons traveling faster than the speed of light, is not a possibility.

In view of what has been presented so far, Transcendent Physics posits that *it is very likely that the information travels via consciousness*. Since consciousness is massless and not transmitted by the electromagnetic spectrum, this would not violate the Special Theory of Relativity. In a moment of truth, some physicists will tell you that there is every reason to believe that the same kind of effect would occur for things larger than electrons, including human beings.

- **Teleportation:** *A fascinating and proven consequence of quantum entanglement is that essential information contained within one of two entangled entities may be transmitted instantaneously, even over a distance of millions of light-years, materializing in the form of a third identical entity.*

Using a photon as an example, the initial photon simultaneously vanishes in the teleportation experiment, as if moving from here through nothingness and showing up there as the third photon. Teleportation has been experimentally demonstrated for electrons, atoms, and molecules. This can be explained if the mode of entanglement is via consciousness.

Photon teleportation is currently being explored as a practical means for advanced quantum computers and cybersecurity systems.

- **Quantum Duality:** *In the quantum realm, the measurement of an entity's properties can reveal either particle-like or wavelike characteristics, depending on the experimental approach.* For instance, a "particle" experiment will yield results consistent with discrete particles, while a "wave" experiment will showcase the entity behaving like a wave. This intriguing duality extends from subatomic particles, such as electrons and photons, to macroscopic entities, including human beings. However, conducting experiments to demonstrate this phenomenon at a larger scale remains a challenge.

  A classic example is Thomas Young's double-slit experiment, replicated with single electrons fired one at a time at two narrow slits facing a detection screen. Upon examining the detection screen after a period of operation, a clear wave-diffraction pattern emerges, characteristic of wave behavior.

  But when detectors are introduced to observe which slit the electrons pass through, the result changes: The screen displays only two distinct spots, revealing the electrons behaving as single particles passing through one of the slits and hitting the detection screen at a single spot opposite one of the slits. In essence, the act of observation alters their behavior, transforming wavelike behavior into particle-like outcomes.

Important Note: Wave-particle duality should not be confused with the nonduality function of the universe, as defined in Reflection 9, Conjecturism 4 of Part I: *Our Immanent, Transcendent, Nondualistic Universe.*

- **Two Places at Once:** *Quantum mechanics suggests that a single entity—be it an electron, a photon, or an atom—can exist simultaneously in multiple states.* If two boxes and an atom are present within a quantum system, prior to observation, the atom can be thought of as residing in both boxes at once. It is only upon opening one box that we observe that the wave function has collapsed, allowing us to locate the entity in one box or the other.

  The probability of finding the entity in a specific box can be calculated from the square of the relevant wave function (ψ^2), which indicates the likelihood based on the "waviness" of the wave function for conditions in that box. The "wavy" nature of the wave function correlates with the probability of *our observation* finding it in a specific box, rather than the object's actual presence there. This concept can be bewildering but has been substantiated by repeated experimentation, encapsulated in the quantum mechanical concept known as the *Principle of Superposition.*

- **Big Is Like Small:** *Although quantum mechanics is essential for describing the state of small entities like atoms and electrons, its principles apply broadly, encompassing everything from individual particles to the cosmos.* In fact, cosmologists grapple with formulating a wave equation that describes the entire universe, which is comprised of these same micro-entities that obey quantum rules.[91] The realization of these effects becomes less apparent as we scale up, yet even at larger dimensions, we continue to create our own realities. This intriguing phenomenon is sometimes referred to as the *Quantum Enigma.*[92]

- **Uncertainty Principle:** Nobel laureate Werner Heisenberg posited that small objects, like electrons and atoms, do not present themselves as "real" entities in our conventional understanding. Instead, they exist as "potentialities," with countless possibilities, of which one possibility is physically manifested as the wave function collapses during observation.

  Heisenberg mathematically demonstrated that merely observing a system disturbs it, frustrating attempts to disprove quantum mechanics' assertion that observation creates the properties of what is observed. To clarify further, *knowing the exact position of a moving particle—say, an electron—necessarily limits our ability to know its velocity, and consequently, its energy, since* $E = \frac{1}{2} mv^2$. Conversely, if we pinpoint a particle's velocity accurately, our knowledge of its position and energy diminishes.

  This interplay of measurement arises from the necessity of employing high-energy photons, which, due to their minuscule size, perturb both position and velocity during observation. While this principle is also true for larger objects, the resultant uncertainties become so minuscule that they remain imperceptible. Consequently, we favor classical Newtonian mechanics—an approximation of quantum mechanics—for tracking larger bodies like spaceships or planets, facilitating accurate predictions. See the example in Reflection 17, calculated for an electron versus a baseball.

- **Quantum Jumps:** Within an atom, electrons spiral around the nucleus at varying energy levels or orbits. Stimulating an electron may cause it to leap from one orbit to another—a transition known as a *quantum jump*. Notably, *while an electron can be observed in its respective orbits before and after a transition, it is forbidden by quantum mechanics from existing in any intermediary state during the femtosecond of its jump to another level of energy.* Therefore, it cannot be observed in between levels.

This perplexing notion troubled both Einstein and Schrödinger, prominent figures in the development of quantum mechanics. Though Einstein contributed significantly to the field, particularly through his elucidation of the photoelectric effect, he dedicated much of his later career to disproving the implications inherent in quantum jumps, finding them philosophically unsettling. He was not successful in his pursuit.

- **Electron Promiscuity:** Electrons surrounding an atom spend more than 95 percent of their time in proximity to the nucleus. However, there exists a tiny probability—infinitesimally small—that these electrons could be located anywhere in the universe. This phenomenon implies that, on a fundamental level, we are all sharing our electrons with each other. *Every atom's electrons are, to varying degrees, intermingled with other atoms throughout the universe, creating an intrinsic connection between all matter.*

The principles mentioned above have undergone extensive verification, consistently demonstrating their accuracy. If you find these concepts perplexing or unsettling, take solace in knowing that you are not alone. Many highly experienced quantum physicists grapple with these conclusions, often steering clear of their implications due to a deep-seated commitment to the Scientific Method and to maintaining the impenetrable boundaries between science, philosophy, and what they perceive as metaphysics. This division became starkly apparent in Europe during the Inquisitional period of the Middle Ages, marking a turning point in the evolution of the Scientific Method as physics, and science in general, emerged distinct from philosophy and metaphysics.

Nevertheless, as research progresses in groundbreaking fields like quantum computing and quantum cryptography, an encounter with consciousness not only becomes inevitable, but necessary. Remarkably, the technical underpinnings of these new domains are inextricably linked to the unusual properties and phenomena we've outlined. This emerging intersection marks the nascent entry into Transcendent Physics, which uniquely and synergistically combines key elements of classical physics with conscious quantum physics and metaphysics.

In this exploration, we are not merely dissecting the mechanics of quantum phenomena; we are also challenging and expanding our notions of reality itself. As we move forward in this journey through Transcendent Physics, we will unravel the layers of complexity that connect the quantum world with our lived experiences, bridging the gap between advanced scientific inquiry and the tangible aspects of consciousness that shape our perception of existence.

Ultimately, this endeavor invites us to reconsider the fundamental nature of reality and the important implications of our conscious engagement with it. This requires merging certain fundamental principles of physics with metaphysics.

To further elucidate these conclusions, it will be beneficial to reflect on the historical trajectory of quantum mechanics, contextualized in a simplified, nonmathematical framework. This will allow us to better grasp the origins and implications of quantum theories, laying the groundwork for a deeper understanding of the relationship between consciousness and the underlying fabric of reality.

REFLECTION 12

A Brief History of Quantum Mechanics

"It is safe to say that nobody understands quantum mechanics."[93]

—Richard Feynman,
Nobel laureate and creator of quantum electrodynamics

The Double-Slit Experiment

Our journey into the realm of quantum mechanics begins in 1801 when the English polymath Thomas Young conclusively demonstrated that light behaves like waves rather than particles, the latter a notion previously suggested by Isaac Newton. Although the wave nature of light was speculated about in the 17th century by Dutch scientist Christiaan Huygens, it was Young who provided definitive proof through his experiments involving what are known in physics as *interference phenomena*.

Young discovered that when a coherent circular wavefront of sunlight was passed through two closely adjacent, vertically parallel, thin slits in a metal sheet, it produced alternating dark and light concentric rings on a downstream projection screen. As illustrated in Figure 1, these results are easily explained if light is wavelike in nature.

The results are similar to those for the interaction of circular waves created in a stagnant pond by simultaneously tossing in two stones that land in proximity to each other. The circular waves created by each stone interact as they travel toward each other.

In Young's experiment, if light waves arrived at a point on the screen in phase—that is, a peak of high amplitude or intensity that aligned with another peak of high amplitude, and a valley of zero amplitude that aligned with another valley of zero amplitude, they combined and produced a wave with larger amplitude. This resulted in *constructive interference*, creating a bright fringe on the detection screen.

If a light wave arrived at a point on the screen out of phase—that is, a peak of a high amplitude or intensity aligned with an equally high-amplitude wave that is in opposite phase (antiphase), meaning the positive displacement (crest) of the first wave met the negative displacement (trough) of the second wave, they combined to cancel each other out to zero amplitude. This resulted in *destructive interference*, creating a dark fringe on the screen. We refer to these projections as *interference rings* or *interference fringes*.

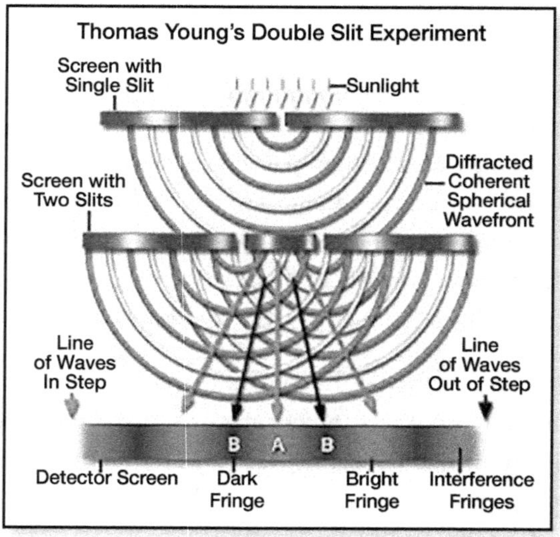

Figure 1: Thomas Young's Famous Double-Slit Experiment

Faraday and Maxwell Clarify the Electromagnetic Spectrum

As the 19th century progressed, brilliant self-taught British chemist and physicist Michael Faraday demonstrated that when the intensity or strength of an electric current changes over time as it flows through a wire in a given direction, it generates a magnetic field perpendicular to the wire—that is, to the direction of current flow. Conversely, when the strength of a magnetic field in a specific direction fluctuates, it produces an electric field and current perpendicular to the direction of the magnetic field. This interplay inspired the mathematical genius of James Clerk Maxwell to synthesize Faraday's findings into Maxwell's Equations—four of the most significant equations in classical physics. These equations elucidate nearly everything about the properties, behavior, and interplay of electricity and magnetism.

The Maxwell equation [$\nabla \times B = \mu_0 j + 1/\mathbf{c}^2\, \partial E/\partial t$] that quantifies how a changing electric field [$\partial E/\partial t$] creates a magnetic field [$\nabla \times B$], contains a constant, "**c**," which is the velocity of the electric field—that is, the velocity of the electromagnetic waves. This allowed Maxwell to calculate the velocity of these waves. Remarkably, the value of "**c**" aligns perfectly with the speed of light, approximately 310,000 kilometers per second or 186,000 miles per second, establishing that both light and all electromagnetic radiation are transmitted as waves at the speed of light. Thus, "**c**" is not arbitrary but is a fundamental constant of the universe.

The electromagnetic radiation spectrum ranges from long wavelengths, low vibrational frequencies, and low energies—like radio waves—to short wavelengths, high frequencies, and high energies—such as X-rays, as shown in Figure 2 on the next page.

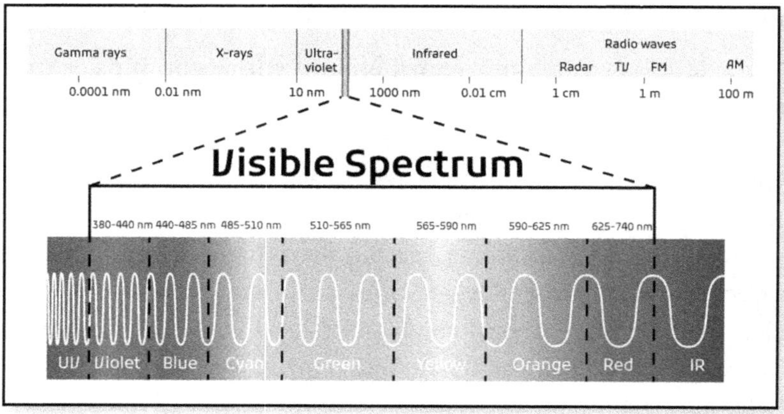

Figure 2: Electromagnetic Spectrum

nm = nanometer, i.e., 10^{-9} meters or 1-billionth of a meter, m = meter

While scientific history often marks the beginning of quantum mechanics at the dawn of the 20th century with Max Planck's discovery regarding the quantum nature of photons, its deepest roots lie quietly subdued in the mid-19th century, when scientists began exploring the connections between the electromagnetic spectrum and blackbody radiation.

Blackbody Radiation Launches the Quantum

The genesis of quantum theory can be traced back to 1859, when Gustav Kirchhoff developed an elementary theory regarding blackbody thermal radiation. A blackbody is an idealized substance that absorbs all electromagnetic radiation that falls upon it and does not reflect any back—for example, a black container. Consequently, when observing a blackbody across the visible spectrum, it appears black since it emits no light—that is, a pitch-black container. However, when heated to a higher temperature, it emits thermal radiation characteristic of that temperature.

Kirchhoff correctly assumed that thermal radiation was wavelike in nature, showing that the energy of emitted blackbody radiation depends on the temperature (T) of the object and the vibrational frequency (ν) of the radiation. The vibrational frequency (ν) is inversely proportional to its wavelength (λ), representing how many times per second a wave vibrates from its peak to its low point or valley. For electromagnetic radiation, the vibrational frequency (ν) is calculated precisely by dividing the speed of light (c) by the wavelength (λ) of the specific radiation: $\nu = c/\lambda$.

Twenty years later, Josef Stefan conducted experiments demonstrating that the total energy emitted by a hot blackbody is proportional to the fourth power of its temperature (T^4). Ludwig Boltzmann subsequently confirmed this relationship theoretically, combining elements of thermodynamics with Maxwell's theory of electromagnetic radiation. While this Stefan-Boltzmann law was significant, it did not sufficiently account for all frequencies or wavelengths of thermal radiation. In 1896, Wilhelm Wien proposed a mathematical relationship that worked for shorter wavelengths, but Heinrich Rubens later showed that it failed for longer wavelengths.

Kirchhoff, who was a professor at Heidelberg University, was offered a prestigious position in Berlin. Initially, his post was offered to renowned physicists Ludwig Boltzmann and Heinrich Hertz, both of whom declined. The position ultimately went to a relatively unknown physicist, Max Karl Ernst Ludwig Planck, who accepted it despite the advice of a colleague and friend, Philipp von Jolly, who had told Planck, "In this field, almost everything is already discovered, and all that remains is to fill a few holes."

On October 7, 1900, Heinrich Rubens, one of Planck's former colleagues, visited him and shared the data and his insights from Wien's work. Several hours after Rubens departed, Planck intuitively deduced the correct formula to explain the radiation emitted from blackbodies *across all wavelengths*.

Max Planck Fathers Quantum Mechanics

However, Planck was not satisfied with merely having guessed the answer; he sought a theoretical foundation for the formula. Despite numerous approaches that proved unsuccessful, he experienced a moment of sheer emotional exhaustion, leading him to make an unprecedented—

some at the time would call it absurd—assumption that would forever alter the landscape of physics.

Planck proposed that the total energy emitted by a blackbody source consists of indistinguishable packets of energy, which he called "quanta," and which were later renamed "photons." This revelation established that the radiation emitted from blackbodies was not a continuous stream of energy, as classical physics had predicted. It was emitted discontinuously as discrete packets of radiation—that is, photons.

Planck postulated that the energy (E) of these quanta or photons is given by the deceptively simple formula: $E = nh\nu$, where n is a positive integer (1, 2, 3 . . .), h is a constant, and ν is the vibrational frequency of the quanta. He could have proposed that electromagnetic energy is directly proportional to frequency, yielding the formula $E = h\nu$.

However, this did not suffice. He needed the positive integer variable of "n," implying that energy is released from the blackbody in discrete packets rather than a continuous flow, a notion that diverged sharply from Newtonian physics. The fundamental and creative leap here is that the energy can change only by integral amounts as packets of energy one at a time, not as a continuous source of energy as permitted by Newtonian physics. So no energy was permissible for, say, $n = 1.5$, or any other non-integer.

Planck speculated that perhaps these quanta were emitted from the vibrating surface structure of the blackbody. He later humbly admitted that his assumption of quanta, which would earn him the Nobel Prize in physics, "was an act of desperation."

Years later, a young patent examiner named Albert Einstein entered the scene. Unable to secure a university position due to perceptions of his inexperience and lack of brilliance, he found solace in his role at the patent office, which afforded him the time to think and imagine—his natural inclination since childhood. In 1905, often termed his *annus mirabilis*, or "miracle year," Einstein published three papers that would revolutionize physics.

Einstein Revolutionizes Physics

Drawing on his knowledge of Planck's postulate of photons, one of these papers addressed the photoelectric effect, which describes the release of electrons from metal surfaces when bombarded by photons of light. This paper catalyzed the rapid development of modern quantum mechanics. Of the other two papers, one focused on the Special Theory of Relativity, and the other provided a mathematical explanation for Brownian motion in liquids, conclusively demonstrating the existence of molecules, which had previously been speculative. These three groundbreaking papers emerged from an "inexperienced, not-so-smart physicist"!

Maxwell's theory of electromagnetic radiation, positing that light is a wave, conflicted with Planck's insight on blackbody radiation and with Einstein's interpretation of the photoelectric effect, which indicated that light consists of particles—that is, photons. Metals contain electrons that can move within their atomic structure, allowing electric currents to propagate through wires. However, these electrons lack sufficient energy to escape from the metal surface. In the photoelectric effect, photons with adequate energy, typically in the ultraviolet range, strike electrons within the metal surface, transferring enough energy to enable them to overcome the force—known as the work function, represented by the Greek letter, phi (ϕ)—that holds them within the metal. This is analogous to a fast-moving marble (the photon) colliding with a slower-moving one (the electron).

According to Maxwell and classical physics, electrons would be excited by ultraviolet photons, and some might gain enough energy to escape from the metal surface. Thus, one would expect the degree of the photoelectric effect to correlate with the intensity (brightness) of the ultraviolet light beam—the brighter the beam, the higher its energy. Classical physics would predict that to generate higher-energy electrons, one could simply shine a brighter ultraviolet light on the metal surface. However, *Einstein found that below a certain critical frequency of the light beam, no electrons are emitted, regardless of the light's intensity (brightness) at that frequency. Furthermore, above this critical frequency, even a very weak light beam (low brightness) can eject electrons.*

Einstein, aware of Planck's work that assumed light was delivered in quanta, made the same assumption in his analysis of the photoelectric effect and derived an equation that accurately described the relationship between the energy of the emitted electron and the frequency of the incident light radiation. His mathematical expression included Planck's constant (h), *now recognized as a fundamental constant of our universe*, valued at 6.63×10^{-34} joule-second.

In view of these findings, physics faced a deep crisis. How could the foundational discoveries of the 19th century, which relied on the wave nature of light, coexist with the particle perspective introduced by Planck and accurately demonstrated by Einstein? The resolution is that both views hold truth, forming the bedrock of a fundamental concept in quantum mechanics known as *wave-particle duality.*

REFLECTION 13

Am I a Wave or a Particle?

"The entire 50 years of pondering have not brought me closer to an answer to the question, 'What are light quanta [photons]?' Today, every Tom, Dick, and Harry believes that he knows, but deceives himself."[94]

—Albert Einstein at age 72 (1951)

Double-Slit Experiments Revisited

Consider the following experiments, which have been conducted many times by numerous investigators with consistent results.

Let's revisit Thomas Young's double-slit experiment, where light behaves as a wave. In our setup, we label the two slits Slit A and Slit B. If we fire very small particles such as mini-marbles at both of these slits, they predictably land at the screen behind either Slit A or Slit B, forming two distinct impact zones.

Now, let's replace the light source normally used in the Young double-slit experiment with an electron gun, capable of firing electrons one at a time toward the slits (see Figure 3 on the next page). Beyond the slits, we place a detection screen, such as a photographic plate, where each electron's impact is recorded as a dark spot.

When electrons are fired individually, each leaves a single mark upon arrival. At first glance, this suggests particle-like behavior. However, after many electrons are fired and their mark accumulates on the screen, the familiar interference pattern emerges, characteristic of waves. This suggests that electrons behave not just as particles, but also as waves! Actual experimental data are shown in Appendix I.

The Mystery Deepens

This experiment challenges classical intuition. If each electron travels independently, how does an interference pattern emerge? One possibility is that each electron somehow passes through both slits simultaneously and interferes with itself. Physicists were deeply perplexed.

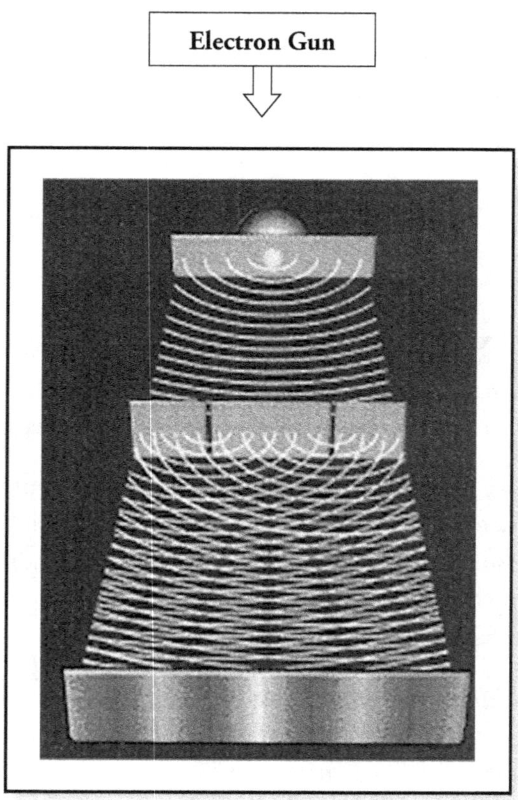

Figure 3: Electron Bombardment of Double Slits

To investigate further, physicists placed behind each slit a special detector capable of "seeing" the transit of each electron to determine which slit each electron passed through. The results were astonishing. With the detectors in place, the interference pattern vanished, and electrons behaved solely as particles, forming two distinct individual bands on the detection screen. It was as if the electrons "knew" that they were being observed and adjusted their behavior accordingly to that of particles!

Richard Feynman, who likely understood the essence of quantum mechanics as well as anyone, once said, "The double-slit experiment has in it, the very heart of quantum mechanics, but in reality, it contains the only mystery."[95]

The Quantum Reality

This phenomenon illustrates three fundamental principles of quantum mechanics:

1. **The Limits of Human Perception:** Our classically programmed intuition fails us in the quantum realm. An electron can exist in a superposition of multiple possibilities, including the possibility of passing through both slits at once.

2. **Quantum Mechanics Is Probabilistic:** Unlike classical physics, which is deterministic, quantum mechanics operates on probabilities. We can calculate the likelihood of an electron being at a certain position, but not its exact trajectory.

3. **Observation Creates Reality:** If we observe which slit the electron passes through, it behaves like a particle. If we don't do this observation, an interference pattern emerges, revealing its wave-like nature. The behavior of the electron depends on how we choose to observe it. This is at the heart of the *Observation Effect* and the *Wave-Particle Duality Concept.*

In 1926, Max Born formalized this probabilistic interpretation, showing that the wave function (ψ) describes a system's probability distribution, with ψ^2 representing the likelihood of various outcomes. For this, he was awarded the 1954 Nobel Prize in physics.

Schrödinger's Dilemma

Despite these breakthroughs, many physicists, including Schrödinger and Einstein, struggled with the implications of quantum mechanics. Schrödinger, skeptical of quantum superposition, proposed his famous thought experiment: A cat is placed in a sealed box containing a mechanism that can release a toxin at an unknown random point in time. Before opening the box at any point in time, according to the principles of quantum probability and quantum superposition, the cat should exist in a superposition of both live and dead states until the box is opened and it is observed, at which time, the wave function collapses and gives a single state—a cat that's either dead or alive.

Today, researchers at the Max Planck Institute of Quantum Optics are testing these ideas with viruses.[96] By using laser pulses, they aim to place a virus in a quantum superposition of two energy states. If successful, this could extend quantum mechanics to living biological systems, potentially reshaping our understanding of life itself.

Philosophical Implications

The double-slit experiment is not merely an abstract scientific demonstration; it has deep philosophical implications. It suggests that reality is not a fixed entity but something that emerges based on observation. If an electron can exist as both a wave and a particle, depending on how it's measured, what does that say about the nature of existence itself?

One of the most perplexing consequences of quantum mechanics is the idea that consciousness itself may play a role in shaping reality. The Observer Effect implies that measurement collapses the wave function, forcing a particular outcome to emerge from a large set of probabilities. Some interpretations—for example, Transcendent Physics, suggest that this means consciousness actively participates in the creation of reality. Others argue that observation is merely an interaction with the system, requiring no conscious agency.

This debate has led to various interpretations of quantum mechanics, including the Copenhagen Interpretation,[97] the Many-Worlds Interpretation,[98] and the Pilot-Wave Theory.[99] Each attempts to explain the fundamental paradox of wave-particle duality in different ways, yet none has been universally accepted. The Uncertainty Principle, formulated by Heisenberg, adds to this mystery by stating that we can never simultaneously know both the exact position and momentum of a particle. This uncertainty is not a limitation of measurement, but an intrinsic property of nature. Consciousness?

Bridging Quantum Mechanics and Consciousness

The relationship between quantum physics and consciousness remains one of the most controversial and intriguing topics in modern science. Some researchers, such as physicist Roger Penrose and anesthesiologist Stuart Hameroff, have proposed that quantum processes within the brain might be responsible for consciousness. Their theory, known as Orchestrated Objective Reduction (Orch-OR),[100] suggests that microtubules within neurons could act as quantum computing elements, potentially linking cognition to fundamental physics.

While this idea remains speculative, it aligns with the broader theme of quantum mechanics challenging our traditional notions of determinism, causality, and the nature of existence. If reality at its most fundamental level is probabilistic, what does that mean for concepts like free will, identity, and the nature of thought?

The late physicist and philosopher David Bohm is someone who thought "out-of-the-box," perhaps more than most of his contemporaries. In his classic work, *Wholeness and the Implicate Order*,[101] Bohm develops a theory of quantum physics that treats the totality of existence in the universe, including energy, matter, and consciousness, as an unbroken whole—similar to some of the principles of panpsychism and Transcendent Physics.

A New Perspective on Reality

The double-slit experiment remains one of the most important demonstrations of quantum mechanics, challenging our assumptions about reality. It suggests that at the most fundamental level, reality is not fixed, but shaped by conscious observation. This realization invites us to rethink not only the nature of matter but also our role in the unfolding of the world—and for that matter, the universe.

As technology advances, experiments continue to push the boundaries of what we know about quantum mechanics. The study of macroscopic quantum phenomena, quantum computing, and quantum biology may eventually provide deeper insights into the fundamental nature of reality. Until then, the question remains: Are we waves, particles, or something beyond both? Continued research efforts in Transcendent Physics may well provide the answer.

REFLECTION 14

The Atom: What Am I?

"In every atom, there are worlds within worlds."

—*Yoga Vasistha*, Ancient Sanskrit text

The Universe Creation

It is well established that our universe was born approximately 13.8 billion years ago in an event known as the Big Bang. Without delving into the intricacies of nuclear and particle physics, we can trace the formation of the first atoms to the aftermath of this cosmic explosive expansion. Initially, after newly born subatomic particles combined, primordial space contained nuclei of hydrogen, helium, and some lithium atoms.

Over billions of years, stars eventually formed by the gravitational attraction of these atoms to each other. Eventually, in a process called nucleosynthesis,[102] nuclear fusion within the high-pressure, high-temperature core of these stars created the heavier elements from the hydrogen, helium, and lithium atoms.

In their final moments, many highly compacted, dense stars underwent supernova explosions, scattering these elements throughout the cosmos. This stellar alchemy produced the 92 naturally occurring elements that make up our universe and everything within it. It is not mere poetry but a fundamental truth of cosmology that we are, quite literally, made from stardust.

Today, our best estimate of the universe's composition reveals a striking fact: Only 5 percent of it consists of the visible matter we perceive with our senses. The rest is an enigma—70 percent is dark energy, and 25 percent is dark matter. Dark energy is not a tangible substance but rather an unseen force driving the accelerated expansion of the universe. Dark matter, though invisible, plays a crucial role in holding galaxies together and stably in place through its gravitational influence. Without these mysterious components, the universe as we know it would not exist.

The Evolution of Atomic Theory

In 1897, J. J. Thomson discovered the electron, but the structure of the atom remained unclear. At the time, the dominant model was the "plum pudding" model, which envisioned neutral atoms as diffuse clouds of positive charge embedded with a balance of negatively charged electrons, much like plums in a pudding. This model, however, was soon upended.

In 1911, Ernest Rutherford conducted a pivotal experiment by accelerating positively charged helium nuclei at a thin sheet of gold foil. While most particles passed through it, a few deflected sharply—some even bounced straight back, as if encountering a solid wall. This unexpected result led Rutherford to propose a radically new model: Atoms consist of a dense, positively charged nucleus surrounded by orbiting electrons, much like planets circling the sun. This was a significant leap forward in our understanding of atomic structure.

However, there was a big problem. According to classical Newtonian mechanics and Maxwell's electromagnetic theory, a particle undergoing circular motion has a continuously changing velocity. A particle with a continuously changing velocity is continuously being accelerated. Therefore, the orbiting electron feels an acceleration, and according to classical mechanics and electromagnetic theory, it must lose energy—radiate it away—and therefore should spiral rapidly into the nucleus.

This implied that all atoms should be inherently unstable—clearly contradicting reality. The resolution to this paradox lay in the emerging field of quantum mechanics, which revealed that electrons do not orbit the nucleus in classical paths but exist in quantized stable energy states, fundamentally altering our view of matter itself. This model was proposed by Niels Bohr, who posited two assumptions leading to the creation of what has become known as the "Old Quantum Mechanics." These assumptions and the Bohr model are discussed in further detail in Reflection 15.

The Atom: A Paradox of Solidity and Emptiness

Though we often speak of atoms, protons, neutrons, and electrons as particles, they are not solid entities in the way we typically conceive of matter. Instead, they are better understood as primarily manifestations of energy and information. The so-called solid objects around us are composed of atoms that are, in reality, mostly empty space.

To illustrate this, consider that over 99.9 percent of your body mass consists of atomic nuclei, which themselves occupy an infinitesimally small fraction of space. The protons and neutrons inside these nuclei are vibrating at roughly one-third the speed of light (~100,000 km/s or 62,000 mi/s). If we could somehow remove all the empty space within our atoms and compress the fundamental particles together, the combined matter of all current 8 billion people on Earth would condense into an object no larger than a common sugar cube—yet it would weigh an astonishing 606 million metric tons.

So, if atoms are mostly empty, why do we perceive objects as solid? The answer lies in quantum mechanics. In the Pauli Exclusion Principle, discussed in Reflection 3, powerful quantum forces prevent electrons from occupying the same quantum state, creating a fundamental and strong resistance to compression. Additionally, electrostatic forces between atoms generate a repulsion that gives rise to the sensation of solidity when we touch an object. In other words, what we experience as "solid matter" is actually an intricate dance of energy fields and repulsive quantum interactions—and as a result, mostly empty space!

The Unseen World within the Atom

The deeper we investigate the atom, the more we uncover a reality that defies common intuition. Protons and neutrons, once thought to be the fundamental building blocks of matter, are themselves composed of even smaller constituents known as *quarks*. These quarks interact via the strong nuclear force, one of the four fundamental forces of nature,[103] which binds them together with an almost incomprehensible strength. Without this force, atoms—and consequently, all matter—would disintegrate.

Even more intriguing is the role of quantum fluctuations within the atom. In the quantum realm, particles do not exist in definite locations but rather as probabilistic wave functions, existing in multiple states simultaneously until observed. This is famously demonstrated in the double-slit experiment, where electrons behave as both particles and waves, depending on whether or not they are being measured. The act of observation itself influences their behavior, raising challenging questions about the nature of reality and the potential role of consciousness in shaping the physical world.

Moreover, Heisenberg's Uncertainty Principle dictates that there is a fundamental limit to how precisely we can simultaneously know both the position and momentum of a particle. This inherent uncertainty at the quantum level suggests that reality is not a fixed, deterministic structure but a dynamic interplay of possibilities. The implications of this principle extend beyond physics, influencing fields as diverse as neuroscience, artificial intelligence, and even theories of free will.

Atoms and Consciousness

Given that the very essence of the atom is nonmaterial—comprised of energy, probability, and information—it is worth considering how this framework relates to consciousness itself. If matter at its most fundamental level is defined not by solid particles but by energetic interactions and wave functions, some scientists and philosophers have asked, "Could consciousness arise from similar principles?"

Transcendent Physics thinks not. It posits that consciousness is a fundamental *intrinsic* aspect of the universe, rather than an emergent property of complex biological systems. Quantum mechanics introduces the possibility that the observer plays a key role in defining reality, hinting at a deep connection between awareness and the structure of the physical world. While these ideas remain speculative, they challenge the long-standing notion that consciousness is merely a by-product of neural activity.

From the vastness of the cosmos to the minuscule structure of the atom, we find a recurring theme: Reality is far more intricate and enigmatic than it appears at first glance. If the building blocks of the universe are fundamental fields of energy and information rather than tangible matter, then what does this say about our own existence? Could it be that consciousness, like the atom, is not what it initially seems?

The journey into understanding consciousness must begin with an extensive and perplexing reevaluation of the material world—a world that is far less "solid" than we once believed.

REFLECTION 15

Bohr's Quantum Jump

"A physicist is just an atom's way of looking at itself."

—Niels Bohr, Nobel laureate in physics,
and a founding father of quantum physics

Bohr Intuits Old Quantum Mechanics

Danish theoretical physicist Niels Bohr built on the ideas of Planck, Einstein, and Rutherford, formulating the first effective model of the atom and its quantum properties. In 1913, Bohr, then a postdoctoral student with Rutherford, published his quantum mechanical model of the hydrogen atom based on a revolutionary idea inspired by Max Planck's work on blackbody radiation and Albert Einstein's work on the photoelectric effect. Both of these concepts relied on the quantum notion that certain physical changes in atomic and molecular phenomena occur in abrupt, discrete steps—that is, quantum behavior, not in a continuous fashion as prescribed by classical physics.

For the neutral hydrogen atom, Bohr envisioned a nucleus containing a single positively charged proton and a single negatively charged electron orbiting it in a planetary fashion.[104] *He made two unprecedented assumptions that contradicted classical physics. First, he proposed that the electron in the atom existed only in stationary fixed orbits.* The orbits are assumed to be *quantized*, and as such, although the electron is accelerating around the nucleus, it does not radiate away its energy and spiral into the nucleus as would be prescribed by classical electrodynamics.

Bohr also maintained that although the single electron in the hydrogen atom could be excited to higher energy states where it could circle the nucleus in different and higher energy orbits, it could absorb and release energy only in specific packets—single quanta of photons. It could not absorb or release energy over a continuous energy spectrum, as prescribed by classical Newtonian physics.

A striking consequence of Bohr's assumption was that an electron, when excited from a lower energy level to a higher one, existed first at the lower energy level and then immediately entered the higher energy level, *but never existed in between.* This was a radical departure from classical physics, akin to *Star Trek*'s "Beam us up, Scotty!" phenomenon, more appropriately known as a *quantum jump*. Schrödinger, already deeply disturbed by the probabilities associated with the superposition principle, famously expressed his frustration concerning "those damned quantum jumps!"[105,106,107]

The second of Bohr's assumptions mirrored that which was originally made by Planck for blackbody radiation and by Einstein for the photoelectric effect, where energy was transferred in discrete packets of photons. *Bohr assumed that the angular momentum of the electron orbiting the nucleus was quantized according to the formula:*[108] $L = mvr = nh/2\pi$, where L is the electron's angular momentum, m is its mass, v is its angular velocity around the nucleus, r is the radius of the electron's circular orbit, h is Planck's constant, and n is a positive integer.

Recall that Planck posited a deceptively simple classical physics formula for the photon's energy, $E = h\nu$, where ν is the frequency and E is "quantized" by the formula $E = nh\nu$, where n is an integer. Similarly, by equating the classical physics formula for angular momentum, $L = mvr$, to $nh/2\pi$, Bohr derived specific quantized orbits for the electron—namely, orbits where $n = 1, 2, 3$, and so on. Furthermore, since the angular momentum was quantized, it could not spiral into the nucleus.

Bohr's assumptions were remarkably successful. They avoided the annihilation of the electron by preventing it from spiraling into the nucleus, but more important, his model allowed precise calculations of the hydrogen atom's energy and spectroscopic properties—particularly the experimentally observed spectral lines as measured by a spectrometer when the atom was excited to higher energies. This agreement between theory and experiment lent credibility to Bohr's assumptions, at least for the hydrogen atom.

Electron Quantum Jumps

When an electron absorbs energy, it jumps to a higher energy level. However, it does not remain there indefinitely. It eventually falls to lower energy levels, emitting excess energy in the form of discrete quanta of photons. This process continues until the electron reaches the lowest possible energy state, known as the ground-state, where $n = 1$ in the equation $E = nh\nu$, and then is stabilized in an orbit around the nucleus. This demonstrated that once the electron was in its lowest energy state, it had nowhere else to go. Unlike classical physics predictions that suggested electrons should continuously radiate energy and spiral into the nucleus, Bohr's model provided a stable atomic structure.

Each quantum jump from a higher to a lower energy level results in the emission of discrete photons, which produce the characteristic spectral lines of hydrogen. The sharpness of the hydrogen spectrum was interpreted as a measure of the discreteness that would eventually be recognized as a fundamental characteristic of all quantum processes. Instead of a broad band of energy emission characteristic of a classical energy continuum, quantum processes occur at precise energy levels, yielding distinct spectral lines. For his contributions, Bohr was awarded the 1922 Nobel Prize in physics.

Beyond Bohr: The Evolution of Quantum Theory

While Bohr's model was a crucial stepping stone in understanding atomic structure, it had its limitations. It worked exceptionally well for hydrogen but struggled to accurately describe more complex atoms. The quantum mechanical framework eventually expanded with the introduction of wave mechanics by Schrödinger, Heisenberg's matrix mechanics, and Dirac's relativistic formulation of quantum mechanics. The wave-particle duality of matter, as introduced by de Broglie, suggested that electrons exhibited both particle-like and wavelike properties, leading to the development of the Schrödinger wave-function equation.

Furthermore, Heisenberg's Uncertainty Principle challenged the classical notion of determinism, asserting that one could never simultaneously know an electron's exact position and momentum. This principle marked a major departure from classical physics and laid the groundwork for quantum field theory and modern interpretations of quantum mechanics.

Bohr also contributed to the ongoing discourse on quantum mechanics with his principle of complementarity, which emphasized the necessity of multiple perspectives to fully understand quantum phenomena—namely, that a complete knowledge of phenomena on atomic dimensions requires a description of both wave and particle properties. His debates with Einstein over quantum mechanics' philosophical implications remain legendary, highlighting the philosophical tension between determinism and the probability at the heart of quantum theory.

In retrospect, Bohr's quantum jump was not merely a physical transition but a conceptual leap in our understanding of reality. It ushered in a paradigm shift that continues to shape modern physics, from quantum computing to the study of quantum consciousness. The discreteness of quantum mechanics, once seen as a theoretical oddity, is now recognized as one of the most fundamental aspects of our universe.

Bohr's legacy extends beyond physics; it has influenced philosophy, information theory, and even our understanding of consciousness. The quantum jump, rather than being an esoteric quirk, may hold the key to unlocking deeper insights into the nature of reality itself.

REFLECTION 16

The Matrix

"The problems of language here are really serious. We wish to speak in some way about the structure of the atoms. But we cannot speak about atoms in ordinary language."[109]

—Werner Heisenberg, Nobel laureate, and a founder of quantum mechanics

Bohr's work was a triumph, but it had limitations and was actually the essence of what would come to be known as the "Old Quantum Mechanics." Notwithstanding its great success in calculating the line spectrum for the hydrogen atom, Bohr's model could not be extended to atoms beyond hydrogen. It didn't even work for helium, the next element after hydrogen in the periodic table. Also, Bohr's theory turned out to be inconsistent with Heisenberg's Uncertainty Principle—a fundamental principle of nature.[110]

However, Bohr's work was a valuable staging point that inspired the development of a more complete theory independently by Werner Heisenberg and Erwin Schrödinger. It ushered in development of what would become the "New Quantum Mechanics," the basis for modern quantum physics.

Hay Fever Leads to Quantum Matrix Mechanics

In June 1925, while recuperating from a serious bout of hay fever on the island of Heligoland in the North Sea, Heisenberg, as a means to distract himself from his physical discomfort, developed what is now called *quantum matrix mechanics*, the first successful and useful theory of modern quantum mechanics. His approach was based on the use of matrix algebra, at that time an obscure branch of mathematics.

As a consequence of the use of this mathematical method, he discovered that the motion of the electron was not precisely definable at the quantum level, and in fact that electrons in an atom did not travel in sharply defined orbits at all, as the Bohr model had assumed. Rather, the motion of the electron was smeared about the atom in a cloud.

Although electrons spend the preponderance of their time within proximity of the nucleus, there is a finite and calculable probability that they can be found anywhere in the universe. This follows from a striking mathematical feature of matrices, which Heisenberg used to represent the electron's position and its angular momentum.[111]

However, many matrices are not commutative, as is the case for normal numerical multiplication. For example, 2 x 4 always equals 4 x 2. However, when considering elements A and B within a given matrix, A x B does not necessarily equal B x A. The noncommutative aspect of certain matrix elements in this area of mathematics, concerned, confused, and disturbed Heisenberg.

Serendipitous Discovery of the Uncertainty Principle

However, he persevered, learning the rules for matrix manipulation and finding that the noncommutative rule ultimately led him to discover important physical implications—namely, that *it is impossible to simultaneously determine with total accuracy, both the position and the momentum of the electron.* The more precisely you know the position of the electron, the less precisely you know its momentum, with the converse being true as well. This became the basis for the famous Heisenberg Uncertainty Principle—a serendipitous discovery in which a mathematical model predicted a profound physical and philosophical reality!

Heisenberg showed quantitatively that if we measure properties of an electron such as energy (E) at a given point in time (t), or its location (x) and momentum (p), then the uncertainty in the measurement of each of these variables represented by ΔE, Δt, Δx, and Δp, respectively, is related by the following inequality equations, where h is Planck's constant:

$$\Delta E \cdot \Delta t \geq h/4\pi \quad \Delta x \cdot \Delta p \geq h/4\pi$$

Because the quantity $h/2\pi$ occurs frequently in mathematical expressions of quantum mechanics, these equations are often written as follows:

$$\Delta E \cdot \Delta t \geq \hbar/2 \quad \Delta x \cdot \Delta p \geq \hbar/2$$

where $\hbar = h/2\pi$, and is called "h-bar."

These equations mean that the uncertainty you measure in the position of an electron (Δx) times the uncertainty you measure in its momentum (Δp) will always be equal to or greater than the value of $h/4\pi$ or $\hbar/2$. This result is neither a statement about the inaccuracy of your measurement instruments nor a reflection of the quality of your experimental methods.

It's interesting to note that Planck's constant, *h*, though one of the smallest numbers you might imagine, 6.63×10^{-34} m^2kg/s, is a fundamental constant for the functioning of our universe.

The uncertainty in these measurements arises strictly from the mathematical wave properties inherent in the quantum mechanical description of nature and is elegantly expressed via the mathematics of matrix algebra. Even with perfect instruments and techniques, the uncertainty is inherent in the nature of all material things, since all material things have wave properties associated with them.

A physical picture of the Uncertainty Principle can perhaps be more clearly drawn from the following explanation: Suppose you want to measure the position of an electron spinning about the nucleus. If you use modern instrumentation to measure its location within a small distance in its orbit, let's say within a distance equal to Δx, then the wavelength of the light required to "observe" the electron must be at least as small as Δx. This means that for the electron to be "seen," the wavelength of the light must interact with the electron and thus must be comparable in size; if the wavelength of the light is larger than the electron, it will be transparent, and therefore, unobservable.

However, the wavelength of light has a certain momentum (p), which was determined by de Broglie to be $p = h/\lambda$. Here, h is Planck's constant and λ is the wavelength of the light. When the photon of light interacts with the electron in order to see it, some of the photon's momentum is transferred to the electron, which then leads to a small change in the position of the electron—picture a billiard ball smashing into a second billiard ball at rest. Thus, the very process of looking at the electron leads to an alteration in its momentum, and therefore a change in its location. As a consequence, there is always an uncertainty about the accuracy of precisely locating the position of an electron.

A more amusing function of the Uncertainty Principle is that you can know where a particle such as an electron is precisely located, but not know what it is doing; or you can know what it is precisely doing, but not know where it is. Heisenberg won the Nobel Prize in physics in 1932 for his work on quantum matrix mechanics.

REFLECTION 17

De Broglie and Schrödinger Make Waves

"I do not like quantum mechanics,
and I am sorry I ever had anything to do with it."

—Erwin Schrödinger, Nobel laureate and creator of the Schrödinger wave equation, lamenting the metaphysical and probabilistic nature of quantum mechanics

Physicists Love Differential Equations

Although Heisenberg's quantum matrix mechanics was an incredible advance, the theory was not readily accepted, as it was based on matrix algebra, at the time, an arcane field of mathematics unfamiliar to most theorists. Instead, physicists preferred to deal with differential equations. Differential equations and the calculus for solving them are one of the most powerful tools for making quantitative sense of the material world, and most physicists are steeped in their use.

Schrödinger to the Rescue

A 38-year-old Austrian physicist by the name of Erwin Schrödinger came to the rescue. A handsome man, throughout most of his life

Schrödinger was a noted womanizer. Although it's a matter of colorful debate by his biographers as to how many affairs he had, one of them has a place in history. While staying with a lovely fräulein at a ski lodge in Arosa, Switzerland, he had a "eureka moment" that led to the development of what became the more accepted mathematical approach to quantum mechanics.

A Prince of an Idea

It's a wave theory and completely based on differential equations. Schrödinger decided to follow up on the then-recent bold development by the French aristocrat and physicist Prince Louis de Broglie. De Broglie had shown that all matter—electrons, marbles and baseballs alike—possesses both particle-like and wavelike properties. However, the quantum implications of wavelength only become important for small particles such as electrons, which have long enough wavelengths to be measured. In contrast, a pitched baseball has such a short wavelength that in our perception of reality, it's meaningless to us.

You can see this clearly from the deceptively elegant formula derived by de Broglie: $\lambda = h/p = h/mv$, where the wavelength (λ) of an object equals Planck's constant (h) divided by its momentum (p), which is its mass (m) times its velocity (v). For example, a 0.14 kilogram baseball thrown by a pitcher at 40 meters per second has a wavelength of 1.2×10^{-34} meters—infinitesimally small compared to the size of a baseball, and for all practical purposes, immeasurable and irrelevant for all occurrences with baseballs. However, for an electron, which has a mass of 9×10^{-31} kilograms, traveling at the same velocity, the wavelength calculates to 1.8×10^{-5} meters = 18 microns—a wavelength that is measurable and a relevant value for electron physics research.

De Broglie presented this work in his doctoral thesis in 1924 at the University of Paris. The work was so creative and unbelievable that his adviser, the famous French theorist Paul Langevin, sent the thesis to Einstein for review, who in turn enthusiastically endorsed the genius of de Broglie's findings.

A short time later in 1927, Clinton J. Davisson and his lab assistant, Lester Germer, at Bell Labs in the United States, fired low-energy, slow-moving electrons at a small single crystal of nickel metal. The regularly aligned nickel atoms on the surface of the crystal provided an atomic-scale diffraction grating for the electrons—just as Thomas Young's larger double-slit experiments had done with photons of sunlight—and revealed the telltale interference pattern on a photographic detection plate—demonstrating again that electrons behave as waves.

A few months later, George Thomson in England carried out a similar experiment using gold foil and obtained the same results as Davisson and Germer. In this instance, the regular array of gold atoms also acted as a diffraction grating. It was now definitively proven that electrons not only behaved as particles, but also as waves—this remarkable phenomenon became known as wave-particle duality.

De Broglie won the Nobel Prize in physics in 1929, just five years after completing his doctoral degree, and Davisson and Thomson shared the Nobel Prize in physics in 1937 for their work showing that de Broglie's theory was correct. An interesting backstory is that George Thomson, who proved that electrons behave as waves, was the son of J. J. Thomson, who won the Nobel Prize in physics 31 years earlier for discovering the electron and showing that it behaved as a particle. Both, of course, were correct.

All of de Broglie's work was based on free-moving particles, those with no forces on them. To have a complete theory, it was important to take into account any forces acting on the particle, which of course is the situation in the real world. This is exactly what Erwin Schrödinger succeeded in doing, starting with a differential equation for a typical classical wave and then taking into account the effects of forces on the particle and the implications of both, the kinetic and potential energy fields in which it existed.

The way that Schrödinger went about developing a means to calculate the properties of quantum objects was to draw on an analogy between what was known in classical physics as geometric wave optics. During the 18th century, when many scientists believed that light was wavelike in nature—Newton was an exception—they did not worry about calculating optical phenomena starting with a partial differential wave equation.

As the wavelength of visible light is quite small compared to the large dimensions defining many physical problems—for example, when passing visible light through various lenses, prisms, or reflections from mirrors, it so happens that a whole set of simpler procedures exist in classical optics to calculate theoretical results. It was unnecessary, then, for Schrödinger to start from first principles using a more mathematically challenging wave equation.

Schrödinger recognized that the classical geometric optics methodology was an approximation of a more thorough wave equation calculation. He reasoned that if he were to instead pursue a more comprehensive and precise analysis, he might be able to discover a general equation for the dynamics of a particle like an electron under any conditions.

The equation he discovered is a complex second-order differential equation in space and time, and one of the most fundamental formulas in modern physics. It bears his name: the Schrödinger Wave Equation. He called his approach *quantum wave mechanics* to distinguish it from the quantum matrix mechanics of his professional archrival, Werner Heisenberg. You might say that even theoretical physicists think about branding their ideas and discoveries.

The Golden Age of Physics

The transition from the Old Quantum Mechanics created by Planck, Einstein, de Broglie, Bohr, and others, to the New Quantum Mechanics as a result of work primarily done by Heisenberg and Schrödinger, was an amazing advance in the field of physics. This era is often referred to as the Golden Age of Physics because of the incredible productivity during a short period resulting in the discovery and development of the special and general theories of relativity and quantum mechanics.

This Golden Age began with Planck's work in 1900 and ended with the explosion of the first atomic bomb at the Trinity Test site in New

Mexico on July 16, 1945, followed by the atomic bombing of Hiroshima three weeks later on August 6, and Nagasaki on August 9.

Further developments followed during this period by other great physicists and mathematicians including P. A. M. Dirac, Max Born, and John von Neumann. Dirac derived Planck's empirical formula for blackbody radiation directly from first principles and created a much clearer mathematical structure for quantum mechanics, so clear that it showed the precise equivalence of Heisenberg's matrix mechanics and Schrödinger's wave mechanics. He also combined quantum mechanics with Einstein's Theory of Relativity, a necessity for calculations dealing with subatomic particles that travel at velocities approaching the speed of light. In doing so, he predicted the existence of the positron, the positively charged antiparticle to the negatively charged electron. The positron was discovered a few years later. Dirac received the Nobel Prize in physics in 1933, just seven years after earning his doctoral degree.

Born's Probabilities Were Not Popular—but on the Money!

Max Born, much to the dismay of many highly regarded scientists, including Einstein and Schrödinger, put quantum mechanics on a definitive statistical basis. He showed clearly that if you solve Schrödinger's wave equation for a certain event or system, the probability of your seeing that event occur is exactly equal to the square of the wave function (ψ^2), or modulus, as it's called in mathematics.

Einstein was so disturbed by this result that he told Born, "I don't believe that God would play dice with the universe," to which Niels Bohr retorted, "Einstein, stop telling God what to do!" Several decades later, Stephen Hawking asserted: "Consideration of black holes suggests, not only does God play dice, but He sometimes confuses us by throwing them where they can't be seen."[112]

As with Einstein, and as mentioned previously, Schrödinger was so distressed by the probabilistic nature of quantum mechanics that he often said that had he known about it in the first place, he would never have created the Schrödinger Wave Equation. Born's research, which showed categorically that quantum mechanics can only have a probabilistic interpretation, caused many physicists stress because of its metaphysical implications. Perhaps this may be the reason why more than 20 years passed until he was awarded the Nobel Prize in physics in 1954.

Why were these and numerous other scientists so disturbed by the implications of quantum probabilities? Because they fly in the face of the Scientific Method and put physics face-to-face with metaphysics. That makes scientists very uncomfortable, primarily because they do not have the mathematical machinery and scientific tools to deal with metaphysical phenomena. Furthermore, hundreds of years of separation of physics and philosophy have led to a disdain for mixing these two fields, which is most unfortunate, as elucidation and perhaps quantifying the nature and metrics of consciousness will almost certainly require merging physics and philosophy (metaphysics). Transcendent Physics will likely continue to face this challenge. This state of affairs has sometimes been called the *quantum enigma.*

REFLECTION 18

Quantum Enigma—"Wonderful Copenhagen"

"Einstein, stop telling God what to do!"[113]

—Niels Bohr, Nobel laureate in physics, and a founding father of quantum mechanics

The Copenhagen Interpretation

Niels Bohr, one of the founding fathers of quantum mechanics and a shrewd statesman for the physics community, immediately knew that something had to be done to calm growing concerns as a result of Max Born's research on the probabilistic nature of quantum mechanics. Most physicists were deeply entrenched in Newtonian classical physics, with which they could calculate very precisely the position or energy of a system at any given point in time. They had great difficulty accepting Born's outcome.

Bohr and his colleagues in Denmark developed what became known as the *Copenhagen Interpretation of Quantum Mechanics*, and to this day it is the mantle that most physicists hold high, so as not to face the discomfort of dealing with consciousness and metaphysics.

Primarily developed by Niels Bohr and Werner Heisenberg, it's a collection of ideas that provides a framework for understanding quantum mechanics. It suggests that quantum particles, before measurement, exist in a superposition of all possible states. The act of measurement forces the particle to "choose" one specific state, a process known as the *wave-function collapse*. This interpretation emphasizes the probabilistic nature of quantum mechanics and the role of the observer in determining reality.

In essence, the Copenhagen Interpretation provides a way to reconcile the mathematical formalism of quantum mechanics with experimental observations, particularly in the context of measurement and the probabilistic nature of quantum phenomena. Although it is one of the most accepted interpretations of quantum mechanics, a number of others exist.

As Bruce Rosenblum and Fred Kuttner note in their incisive book, *Quantum Enigma*, "Classical physics explains the world quite well; it's just the details it can't handle. Quantum mechanics handles the details perfectly; it's just the world it can't handle."[114]

Bohr admits in his formulation of the Copenhagen Interpretation that an observation produces the property or system observed. He defines an observation as that which occurs whenever an atomic-scale object interacts with a large-scale object—for example, when a photographic film is hit by an electron and records where the electron landed. Bohr maintains that the film "observes" the electron. He then goes on to say that people never deal directly with objects in the quantum realm, so we don't have to worry about their physical reality or lack thereof.

Actually, the human eye can detect a single photon; it's just that our brain is not wired to process such a weak signal.

Principles of the Copenhagen Interpretation

As Rosenblum and Kuttner summarize, the Copenhagen Interpretation of Quantum Mechanics is based on three fundamental technical principles:[115]

1. **Probability**—The square of the wave function (ψ^2) in a given region of space-time is the probability that the object will be found in that given region. It's a subtle but important point to recognize that *this probability is not where the object is located, but where you will find it.* What this means is that the object was not there until you observed it.

   If we consider the presence of a single atom, associated with two boxes in a quantum system, quantum mechanics tells us that there is no atom in addition to its wave function. Each box contains the atom's wave function, and therefore, we may be unable to fathom the fact that the information and energy associated with each atom is actually superpositioned in both boxes until we observe it, at which time the wave function collapses and the atom is found in only a single box. The wave function is the only way to communicate information in quantum mechanics. The wave function is the entire story.

   Some physicists say that there is no atom in addition to the wave function, and actually, the wave function is synonymous with the atom. I would not go that far, but I believe that the atom is present exclusively as an entity of energy and information and not "materially" or as a second-order partial differential equation. The wave function certainly contains information, but not energy, although it contains information *about* energy. The wave function does not carry specific information as to how a particular entity will behave or a specific event will occur. Rather, it is an expression of potential or statistical probability. This is in stark contrast to classical physics, where answers are always exact, specific, and precisely predictable. This kind of metaphysical behavior startled and concerned scientists when it was first observed.

2. **Uncertainty:** According to Heisenberg's Uncertainty Principle, if we consider the position (x) and momentum (p) [p = mass x velocity] of a small particle such as an atom, the more precisely we know its velocity (v) or momentum (p), the less precisely we know its position, (x) and conversely. Any observation disturbs things

enough to prevent disproof of quantum mechanics' assertion that the act of observation itself creates the thing observed—think of a "viewing" photon crashing into the electron you wish to observe.

3. **Complementarity:** The particle-like and wavelike behaviors of a microscopic object are complementary. A complete description of the object requires that we make use of both behaviors, but never at the same time. This is a statement of the complementarity of classical mechanics and quantum mechanics. The macro world obeys classical mechanics, and the micro world obeys quantum mechanics. If we do a wavelike experiment, we will always get a wavelike answer; if we do a particle-like experiment, we will always get a particle-like answer.

The Copenhagen Interpretation has enabled most theoretical physicists to go about their work without having to confront consciousness and metaphysics. There's less stress, and they sleep better at night.

Additional Considerations

While the Copenhagen Interpretation remains the dominant framework, alternative interpretations challenge its implications. The Many-Worlds Interpretation, proposed by Hugh Everett, eliminates wave-function collapse altogether, suggesting that all possible outcomes coexist in parallel realities (universes). This idea, though mathematically consistent, raises deep philosophical questions about the nature of reality. Furthermore, it is highly unlikely to ever be testable in the laboratory.

In other studies, physicists such as Roger Penrose have explored the intersection of quantum mechanics and consciousness, proposing that quantum effects in the brain might play a role in cognition.

The late David Bohm's pilot wave theory is deterministic and explicitly nonlocal. The velocity of any one particle depends on the value of a guiding equation, which depends on the configuration of all the particles under consideration. It's a complex approach, which has not been extensively further developed by others.

Ultimately, the Copenhagen Interpretation provides a practical tool for physicists, but it leaves a number of mysteries unresolved. What does it truly mean for reality to be observer-dependent? Is consciousness itself entangled with the fabric of quantum mechanics? These questions, once considered purely metaphysical, are increasingly entering the realm of scientific inquiry, especially Transcendent Physics.

REFLECTION 19

Entanglement: I Feel Your Presence

". . . Thou canst not stir a flower without troubling a star."

—"The Mistress of Vision" by Francis Thompson[116]

Spooky Action at a Distance

Entanglement is one of the most enigmatic aspects of quantum mechanics, so much so that Einstein famously derided it as "spooky action at a distance." In essence, entanglement suggests that disturbing a quantum object, such as an atom, can instantaneously—faster than light speed—influence the behavior of another atom, regardless of the distance separating them, even in the absence of any "apparent" physical force or medium. This phenomenon exemplifies the nonlocal nature of quantum mechanics and provides a compelling foundation for the concept of universal connectedness. Metaphysical interpretations often echo this notion, asserting that "we are all connected," referring to the concepts of Immanent, Collective, and Cosmic Consciousness, which are posited by Transcendent Physics to be deeply intertwined throughout the cosmos.

The three defining characteristics of quantum entanglement and nonlocality are:

1. **Instantaneous Interaction:** The effect occurs faster than the speed of light, seemingly defying Einstein's Special Theory of Relativity.

2. **Medium Independence:** No field or medium is required to facilitate the connection, meaning that no known force can obstruct or inhibit it.

3. **Distance Invariance:** The strength of the interaction remains constant regardless of spatial separation, whether across a few meters or across galaxies.

It is no surprise that many physicists find these attributes both perplexing and unsettling. Some researchers exploring the intersection of physics and metaphysics speculate that entanglement might underlie certain paranormal phenomena, such as remote viewing—the purported ability to perceive distant objects or events during heightened energy levels of consciousness. Governments, including those of Russia and the United States, have invested heavily in military research on psychic espionage, seeking to harness such potential abilities.[117] Similarly, some theorists propose that entanglement might explain why individuals occasionally experience an immediate, inexplicable awareness of a distant loved one's distress. While most entanglement experiments involve subatomic particles such as electrons or photons, groundbreaking work has extended the phenomenon to larger systems. In a remarkable experiment reported in *Nature*, J. D. Jost and colleagues[118] successfully entangled two mechanical oscillators—represented by separated ion pairs—demonstrating that altering the vibration of one oscillator instantaneously affected the other. A transcendent physicist would argue that such entanglement arises because all matter—for example, mechanical oscillators—possesses some degree of consciousness, and it is this inherent consciousness that facilitates their connection. They posit that since consciousness is massless, it travels faster than the speed of light—that is, instantaneously.

Reflecting on the paradox of entanglement, Einstein once remarked, "The Lord God is subtle, but malicious He is not." However, after years of unsuccessfully grappling with its implications, he wryly conceded, "I have second thoughts. Maybe God *is* malicious." This reluctant admission underscores the mystery of entanglement—a phenomenon that challenges our deepest assumptions about reality, connection, and the very nature of consciousness itself.

REFLECTION 20

A Bell Tolls for Dr. Bell

"You create your own reality."[119]

—Jane Roberts

Modern cosmology theory maintains that the instantaneous birth of the universe occurred 13.8 billion years ago when "something" explosively expanded from a subatomic singularity—a small entity, billions of times smaller than an atom. Transcendent Physics tells us that "something" was a subatomic singularity containing nothing but pure undifferentiated consciousness, which eventually transformed into all of the matter and energy in the universe.

Therefore, we can say that the "substance"—consciousness—which created all energy and matter in the universe was at one time all "one"—intimately connected in a subatomic singularity at the moment of the Big Bang's initiation—and because of the entanglement principle, all energy and matter throughout the universe continue to be interconnected. We live in a nonlocal, connected universe.

Quantum Masters Were Disturbed

Einstein referred to entanglement as "quantum weirdness." His fear of this "spooky action at a distance" was well founded in that any objects that have ever interacted will continue to instantaneously influence each other, and do so in a time frame faster than the speed of light. How could that possibly be?

Einstein and other notable scientists conjectured that there must be hidden variables in quantum mechanics, factors that, if known, could explain all of this quantum weirdness. But Irish physicist John Stewart Bell, while on a sabbatical, proved that this could never be the case. There are no hidden variables, and quantum weirdness is an intimate part of the universe, despite the fact that it makes some of us uncomfortable—especially physicists.

A Most Important Discovery

Bell developed what became known as Bell's Inequality[120] and published his findings on November 4, 1964, in a journal called *Physics*. The journal was short-lived, but the paper became famous, laying the foundations for the modern field of quantum-information science.

Bell's Inequality has been hailed by some as "the most important discovery in science in the last half of the twentieth century."[121] This Inequality formula made it possible to differentiate between quantum mechanics' correct conclusion concerning the presence of nonlocality throughout the universe and, therefore, the creation of True Reality by our subjective observation, versus any alternative theory involving "secret" instructions or hidden variables.

Bell's work provided a way to test what was previously considered a purely philosophical statement or conjecture. In fact, it opened the door to a new area of pursuit: *experimental philosophy*. All research applying his inequality equation has unequivocally shown that no physical theory using local hidden variables can reproduce quantum mechanics' predictions. Universal connectedness and "spooky action at a distance" are a given. Everything in the universe is connected by entanglement, and according to Transcendent Physics, that connection occurs strictly and directly via consciousness.

Initially known as Bell's theorem, he definitively proved that his mathematical inequality, if violated, would categorically prove that nonlocal, universal connectedness is a cosmic reality. It consists of two assumptions: The *Reality Assumption* conjectures that all objects in our world have physical properties that are not created by observation—it negates the Observer Effect. The *Separability Assumption* posits that two objects can be separated from each other at any distance, such that what happens to one does not influence the other—it negates *nonlocality.*

Using an elegant, relatively simple, and straightforward mathematical derivation, Bell proved that certain measurable, observable physical quantities had to be larger than other well-defined measurable, observable quantities for these two assumptions to be correct. This provided an experimentally testable prediction of Bell's Theorem, which has since become known as Bell's Inequality. In a word, if these assumptions were violated, then nonlocal "spooky action at a distance" was a reality of our universe.

When the appropriate experiments were finally accomplished, first by Alain Aspect and coworkers in France,[122,] and then by many other physicists around the world, Bell's Inequality was shown to be violated in every case. This meant that *reality* and *separability* assumptions were both incorrect for our universe.[123] *Therefore, we create our reality by observation, and all things are connected such that instantaneous entanglement, and therefore, connectivity, is prevalent throughout our nonlocal universe.* For this work, which provided valuable insight into True Reality throughout the universe, Aspect received the 2022 Nobel Prize in physics.[124] Had Bell not died 32 years before this, he surely would have been one of the Nobel recipients.

Distance Is Irrelevant

Experiments have been done over distances of hundreds of kilometers with photons, molecules, and other small molecular systems, and they always verify the veracity of entanglement and Bell's Inequality. In 2017,

Chinese researchers, using one of their satellites in communication with a ground-based observation station, set a record for instantaneous connection—1,203 kilometers, nearly 750 miles.[125] Extrapolation of these results indicates that the results would be the same over a distance of billions of light-years.

Measuring one particle of an entangled pair immediately affects its counterpart, no matter how far apart they are, and it does so instantaneously—that is, faster than the speed of light. Extrapolation of numerous sets of data definitively demonstrates that one particle could just as well be at the other side of the galaxy, and the communication would still be instantaneous.

These effects apply not just to microscopic systems but to large ones as well. Communication between the entangled pair via a signal from the electromagnetic spectrum would be a violation of Einstein's Special Theory of Relativity. The principles of Transcendent Physics posit that this communication occurs by means of massless consciousness and therefore is not a violation of Special Relativity.

In principle, any two objects, no matter what their size and mass are, that have ever interacted are forever entangled—the behavior of one instantaneously and forever influences the other, and vice versa. Indeed, our universe has a mysterious connectedness—consciousness—that goes beyond known classical and quantum forces. Quantum computers of the future are expected to take advantage of the use of entanglement, potentially leading to revolutionary advancements in quantum computation, cryptography, and information processing.

Additionally, some physicists and philosophers speculate that the underlying nature of this connectedness may be key to understanding deeper questions about reality, perception, and even consciousness itself. Could it be that entanglement is not just a physical phenomenon but a fundamental aspect of awareness? Some interpretations of quantum mechanics suggest that consciousness plays a direct role in shaping the nature of reality, an idea that aligns with the principles of Transcendent Physics.

Resistance Going Forward

Bell's Inequality demonstrated his talents as an insightful physicist and, to the dismay of many in the physics community, a metaphysicist, as well. Mixing metaphysics and physics was, and still is, considered *not* respectable science. So it wasn't surprising that when Alain Aspect suggested his experiment, Bell asked him if he had a permanent, tenured faculty position, and if not, was he sure he still wanted to proceed, as it might have a negative impact on his professional future. Even when Aspect shared the 2022 Nobel Prize in physics, the Nobel Committee was somewhat guarded in wording their citation: *For pioneering quantum information science.*

Bell's Inequality was actually more pioneering and profound than proving the non-existence of the *Separability* and *Reality* assumptions. He had developed a wide-ranging *No-Go* theorem,[126] meaning that proof that *any* theory operating under the Separability and Reality assumptions cannot describe the world we live in.[127]

It's heartening and profound that beyond its metaphysical implications, Bell's Inequality has led to revolutionary technologies in quantum cryptography, quantum computing, and quantum teleportation.

The biggest challenge with accepting many of the possibilities for Transcendent Physics is the continued disbelief and detractors among knowledgeable scientists, who persist in trying to disprove Bell's Inequality, nonlocality, the Observer Effect, and other realities of quantum physics, which they consider metaphysical pursuits. In some ways, this resistance is similar to the challenges faced by Galileo, Planck, and other scientists, who changed the trajectory of the physical sciences with their discoveries, leading humanity toward a clearer and deeper understanding of life and the universe.

REFLECTION 21

Teleportation—Beam Us Up, Scotty!

"We are not only observers. We are participators.
In some strange sense, this is a participatory universe."

—John Archibald Wheeler[128]

Another astonishing aspect of quantum entanglement is teleportation—the transmission of quantum information across space instantaneously. As mentioned in Reflection 9, Transcendent Physics posits that matter and energy can be converted into consciousness, teleported, and then reconverted back into matter and energy. What once seemed like science fiction[129] is now an experimental reality.

Sounds Like Science Fiction

Anton Zeilinger and his team at the University of Innsbruck successfully demonstrated quantum teleportation using entangled photons.[130] In their experiment, quantum information from one photon (#1) was instantaneously transferred to another distant photon (#2), effectively re-creating the original photon's quantum state at the new location as a third photon (#3). Importantly, the original photon (#1) ceased to exist in its initial form, reinforcing that what was teleported was not matter itself but its informational essence.

This discovery has important implications for quantum computing and secure communications. IBM scientists and other research teams are actively exploring the applications of quantum teleportation, particularly in ultrafast, superluminal networks. These networks could revolutionize data transfer and encryption, leveraging the fundamental principle of nonlocality—where entangled particles remain correlated regardless of distance. Though we may not yet fully grasp the implications, this is good science, paving the way for breakthroughs beyond our current technological and conceptual barriers.

The success of quantum computing may depend on teleportation and other quantum phenomena that challenge classical notions of space and time. Experiments violating Bell's Inequality suggest that entangled particles exchange information faster than the speed of light, appearing to present a direct challenge to Einstein's theory of special relativity. However, Transcendent Physics offers a resolution: Transport occurs via massless consciousness, which does not violate special relativity, as Einstein's speed limit applies only to the electromagnetic energy spectrum and to objects with mass.

In addition to its technological potential, quantum teleportation raises deep philosophical and scientific questions about the nature of reality. If information, rather than physical substance, is what truly defines an object, then our understanding of matter itself may need revision. Could it be that at the most fundamental level, everything is information, and matter is merely a manifestation of this information in consciousness? This perspective aligns with Transcendent Physics, which posits that reality is structured by underlying informational and consciousness processes.

The Copenhagen Interpretation of Quantum Mechanics acknowledges that the assumptions of separability and realism are violated, yet it maintains that one should "shut up and calculate," avoiding metaphysical implications. However, the significance of entanglement cannot be ignored. Some physicists reluctantly accept that entanglement operates faster than light speed but insist, "Perhaps entanglement travels faster than light, but surely awareness cannot!" This reflects an ongoing resistance to acknowledging consciousness as an *intrinsic* and fundamental aspect of reality, and perhaps, a misunderstanding of the true nature of awareness.

Transcendent Physics, in contrast, asserts that all matter and energy possess some level of consciousness. If teleportation is possible in the quantum realm, then could it be possible at macroscopic scales as well? While mainstream physics has not yet bridged this gap, historical paradigm shifts have taught us that today's impossibilities often become tomorrow's scientific breakthroughs. As quantum computing advances, the evidence for this perspective may become undeniable. Perhaps it will compel us to rethink our understanding of reality—our existence, our consciousness, and our place in the cosmos. The principles underlying this framework are outlined in the Conjecturisms in Reflection 9, providing a foundation for deeper exploration.

A final note: The great physicists of the quantum era—including Planck, Einstein, Bohr, Heisenberg, Dirac, Feynman, and Hawking—all reached two important conclusions: (1) Quantum mechanics remains the only scientific theory that has never made an incorrect prediction, and (2) it is impossible to engage deeply with quantum mechanics without confronting the intrinsic role of consciousness. Could this be a message hidden within the nature of reality?

Just as Newtonian physics is an approximation of quantum mechanics, could quantum mechanics itself be an approximation of Transcendent Physics—a deeper framework where the universe's laws extend beyond our current understanding? Perhaps when we finally grasp these laws, we will arrive at the ultimate Theory of Everything (TOE)—one that unifies not only matter and energy but also the fundamental nature of consciousness itself. If consciousness is *intrinsic* to the fabric of reality, then understanding its role may be the missing key to unlocking the deepest mysteries of the cosmos.

PART III

What Does This Mean for Our Reality?

"The truth in you remains as radiant as a star,
as pure as light, as innocent as love itself.
And you are worthy that your will be done!"

—Jesus Christ (through Helen Schucman)

—*A Course in Miracles*[131]

REFLECTION 22

Quantum Reality

"It is possible in quantum mechanics to sneak quickly across a region which is illegal, energetically."[132]

—Richard Feynman,
Nobel laureate in quantum electrodynamics

The development of Transcendent Physics has progressed over the last decade and will likely continue to do so, albeit at a modest pace. This is primarily due to sociocultural constraints and a long-standing commitment to the principles of the Scientific Method. While the Scientific Method has served humanity well, now and no doubt in the future, engaging deeply with Transcendent Physics inevitably brings scientists face-to-face with metaphysics—even more so than quantum physics. History demonstrates that such encounters often restrict scientific progress.

Crossing this barrier will require scientists prepared to navigate the challenges of securing research funding in a climate skeptical of the methodologies necessary for Transcendent Physics. Although quantum physics may not be as effective as Transcendent Physics in making the conceptual leaps required to understand the full nature and function of consciousness and the universe, it remains an invaluable tool at its interface with Transcendent Physics. Therefore, it is crucial to grasp the mechanics of conscious quantum physics and its bridge to Transcendent Physics.

The Origins of Quantum Insights

Some fundamental aspects of quantum mechanics were not first uncovered through the Scientific Method but were intuited over 2,000 years ago by Eastern Wisdom thinkers and sages. This historical insight offers a compelling clue regarding the next steps necessary to further our understanding of consciousness and its place in the cosmos.

Throughout history, numerous spiritual traditions have described reality as being fundamentally interconnected and influenced by the observer's awareness.[133] These ancient perspectives resonate with modern quantum mechanics, particularly the Observer Effect and quantum entanglement. The Upanishads of ancient India, Buddhist philosophies, and Taoist teachings all reflect an understanding that reality is shaped not only by material interactions but also by the presence and awareness of conscious entities.

While modern quantum theory provides a mathematical and scientific framework suggesting how consciousness might operate within species, systems, and the universe itself, it remains insufficient for unlocking the full power of Transcendent Physics—for example, how does the consciousness-brain-mind triad create, transmit, and translate a thought or the sense of an experience—that is, a sense of overwhelming awe? This is the classic Hard Problem in philosophy and neuroscience.

By analogy, just as Newtonian physics gave way to quantum physics—ushering in a new paradigm—quantum physics is poised for a similar transition to Transcendent Physics. This progression is essential for understanding True Reality and leveraging that knowledge to improve both individual lives and society at large.

Step 1: Internalizing the Interface Between Quantum and Transcendent Physics

Based on the discussions thus far, the following principles summarize the currently known framework of Transcendent Physics:

1. Conditions to create all mass and energy in the universe were set 13.8 billion years ago through the explosive expansion of a singularity—a spherical entity smaller than a subatomic particle.

2. At time zero, the singularity's sole constituent was pure, undifferentiated consciousness. The essence of all consciousness throughout the universe is precisely the same and constant. Its only metric of differentiation is energy level, as measured within an infinite spectrum, completely unrelated to the electromagnetic energy spectrum of classical and quantum physics.

3. During the initial primordial expansion, consciousness generated massless elementary particles such as quarks and gluons. Shortly thereafter, consciousness formed the Higgs energy field,[134] permeating the entire expanding universe and granting mass and consequential gravitational forces to these elementary particles and all matter, including Dark Matter, through a process called *Transubstantiation*.[135]

4. Prior to initiation of the Big Bang, all consciousness and its embedded eternal information were in intimate contact within the singularity. Consequently, all consciousness-derived mass and energy throughout the universe remain eternally connected via the universal presence of quantum entanglement.

5. Subatomic particles combined to form atomic structures, captured by the gravitational force of massive stars. Through nuclear fusion processes called *nucleosynthesis*[136] within the gravitationally intense interior pressures and temperatures of these stars, all 92 naturally occurring elements emerged, and were subsequently distributed throughout the universe via numerous explosive supernovae events.

6. Consciousness follows the same conservation laws as mass and energy—it cannot be created or destroyed—only reversibly converted into its equivalent of energy and/or mass. However, being massless, it moves superluminally, transcending the speed of light.

7. Consciousness can be transformed into mass or energy, and vice versa. This reversible process underlies the cyclical creation, dissolution, and reformation of universes, ensuring the retention of information across cosmic epochs.

8. Only 5 percent of the universe's energy and mass is physically observable. The remaining 95 percent consists of Dark Energy (70 percent) and Dark Matter (25 percent). Transcendent Physicists propose that over billions of years and massive changes in gravitational forces, the universe will eventually enter a contraction phase—the Big Crunch—leading to creation of the next universe—via the Big Bounce.[137] These cosmic cycles are an eternal process.

9. Consciousness manifests in three primary functional forms: Immanent (True Reality in the four-dimensional physical universe), Collective (interconnected consciousness throughout the energy and mass of the cosmos), and Cosmic (an infinite, eternal, Divine energy source—*God,* if one chooses to call It that).

10. The universe exists as an evolving symphony of physical and consciousness evolution, guided by the Six Cosmic Core Values: truth, wisdom, understanding, forgiveness, love, and compassion. These values are not arbitrary or based on any moral injunction, but are essential for unfolding the Meaning and Purpose of the Universe.

11. The Meaning of the Universe is its ongoing transformation in a continuous, synchronous evolution of physicality and consciousness, both of which are mutually and inextricably interdependent. The Purpose of the Universe is the perpetual elevation of the overall energy level of its consciousness.

12. Humans represent the pinnacle of physical and consciousness evolution on Earth.

13. The Meaning of Life is for each species to recognize and embrace its innate strengths and potential. The Purpose of Life is to contribute to the Earth's and universe's eternal physical and consciousness evolution.

14. By accessing Cosmic Consciousness, it is possible to manifest phenomena in our four-dimensional universe, provided they adhere to the laws of manifestation (see Reflection 9, Conjecturism 10).

To further illuminate these principles, we must recall the 11 core tenets of Transcendent Physics outlined in Reflection 11:

1. **The Observer Effect:** Reality is actively manifested and shaped by observation—that is, by consciousness.

2. **Infinite Possibilities:** Reality is based on a domain of an infinite number of potential outcomes. A specific one is created by observation, which causes a collapse of the overall wave equation for all possible outcomes describing the outcome observed.

3. **History Creation:** Observation determines not just the present outcome, but retroactively alters the past for it to be consistent with the observed outcome.

4. **Entanglement:** All energy, mass, and information in the universe are interconnected and instantaneously accessible via Collective Consciousness.

5. **Teleportation:** Information, energy, and matter can be transmitted superluminally by first converting it to consciousness, transmitting it, and then converting it back to its original form.

6. **Quantum Duality:** Matter can appear as either a particle or a wave, depending on the means of observation. This complementary duality is necessary for a complete description of any quantum event.

7. **Two Places at Once:** At the quantum level, particles can simultaneously exist in multiple states. The final stable state is created

by observation as the result of the collapse of the relevant wave function.

8. **Big Is Like Small:** Quantum phenomena scale to macroscopic levels, though they are more difficult to detect at these levels.

9. **Uncertainty Principle:** For certain pairs of fundamental atomic properties (for example, position and momentum), both cannot be simultaneously measured precisely. That is, in two dimensions, the uncertainty in position (Δx) of a particle times its uncertainty in momentum (Δp) always obeys the inequality, $\Delta x \cdot \Delta p \geq \hbar/2$.

10. **Quantum Jumps:** Particles transition between energy states without existing in forbidden intermediate positions between those states.

11. **Particle Promiscuity:** Subatomic particles exhibit a finite probability of appearing anywhere in the universe. The whole is one.

Step 2: Achieving Higher Consciousness to Engage Transcendent Physics

A fundamental requirement for advancing Transcendent Physics—whether in scientific study or personal application—is the attainment of deep, high-energy states of consciousness.[138] This presents a significant challenge, particularly for those entrenched in the conventional Scientific Method. More than theoretical acknowledgment, it requires sustained dedication to deep meditative practices and inner transformation to choose and interpret the most effective paths of research in Transcendent Physics.

To deepen engagement with Transcendent Physics, various disciplines can offer structured methodologies. Numerous meditation techniques, altered states of consciousness through sound frequencies, and controlled psychedelic—psylocibin, LSD, DMT, and 25I-NBOMe—experiences have all been recognized as pathways to accessing higher energy states of awareness.[139] Studies in neuroscience have begun to reveal how such states correlate with altered brain-wave patterns and enhanced cognitive function, providing scientific validation for what mystics and sages have known for millennia.

This is not simply a single exercise of convenience—it is a lifelong commitment to learning and self-actualization. By elevating our consciousness through disciplined practice,[140] we open pathways to accessing and utilizing the principles of Transcendent Physics, ultimately transforming our understanding of True Reality, our deeper *self*, and the universe. The next frontier is not just technological or theoretical—it is the evolution of consciousness itself.

REFLECTION 23

The Architecture of Transcendent Reality

"Your body is not who you are.
The mind and spirit transcend the body."

—Christopher Reeve, actor[141]

The deepest philosophical underpinning of Transcendent Physics is that the fundamental nature of reality is not determined solely by physical laws, but by the presence of consciousness itself.

The scientific journey from classical mechanics to quantum mechanics was, in many ways, a paradigm shift as significant as the transition from a geocentric to a heliocentric model of the solar system, providing a deeper understanding of not only our solar system, but the universe itself. Yet, as profound as quantum physics is, it remains incomplete. It provides a rigorous framework for understanding subatomic particles, probabilities, and to some degree, quantum entanglement and teleportation, but it fails to fully address the nature of consciousness and its role in shaping reality. This shortcoming sets the stage for Transcendent Physics, which aims to bridge the gap between material reality and the fundamental role of consciousness.

To engage with Transcendent Physics, we must first understand the architecture of reality as it is currently perceived. Quantum physics has already exposed cracks in our classical worldview, revealing that reality is not as solid as once believed. Wave-particle duality, quantum superposition, teleportation, and quantum entanglement suggest a deeper substrate of existence—one that is dynamic, nonlocal, and influenced by observation and awareness—that is, by consciousness. These principles hint at an even deeper structure: the *Architecture of Transcendent Reality*.

Quantum Mechanics: The Staging Ground for Transcendent Physics

Despite its revolutionary implications and accomplishments, quantum mechanics has largely been confined to mathematical formulations and experimental validations, leaving many of its philosophical and metaphysical consequences and implications underexplored. The Observer Effect, for instance, suggests that reality is influenced, if not outright determined, by conscious observation. This is a profound insight, one that has yet to be pervasively accepted and integrated fully into mainstream scientific thought.

Likewise, quantum entanglement challenges classical notions of locality, demonstrating that two particles can remain instantaneously connected regardless of distance. Explaining this interconnectivity based on the electromagnetic spectrum of classical or quantum physics violates the Special Theory of Relativity. Transcendent Physics solves this issue by maintaining that the connection-communication is via massless consciousness, which travels superluminally.

Such interconnectedness is not limited to the microscopic world; rather, it provides a compelling hint that consciousness itself may function through a similar nonlocal mechanism. If consciousness can be shown to operate not only in our four-dimensional world but also beyond the constraints of space-time—a primary goal of Transcendent Physics—then the implications for human potential and the understanding of reality are profound.

The Multilayered Fabric of Reality

To conceptualize Transcendent Physics, we must move beyond the conventional four-dimensional model of the universe and consider a more complex, multilayered structure of reality. This architecture consists of three primary layers:

1. **Physical Reality (Immanent Consciousness)**
   - The tangible, measurable world where classical and quantum mechanics govern interactions.
   - Subject to the constraints of space, time, and the energy of the electromagnetic spectrum.
   - Consciousness is nonlocal, tethered to biological organisms and nonbiological entities alike, but not solely confined to their physical geometry.
2. **Collective Reality (Collective Consciousness)**
   - A realm where consciousness operates beyond the individual entity, forming an interconnected network across species, matter, and even across space and time.
   - Quantum entanglement provides a physical analogy for this interconnectedness of the consciousness energy field. As mentioned above, the latter is massless and distinctly different from the electromagnetic energy field; and as such, it is not constrained by the Special Theory of Relativity, and thereby functions by instantaneous connection and communication.
   - The concept of *morphic resonance*, as proposed by Rupert Sheldrake,[142,143] suggests a field-like medium—beyond that of the electromagnetic spectrum—in which collective memory and experience are stored. Transcendent Physics posits that this field is consciousness.

3. **Transcendent-Immanent Reality (Cosmic Consciousness)**

- The ultimate foundation of existence, from which all other layers emerge.
- Not only present in the physical universe, but also beyond the constraints of space, time, and energy.
- The seat of universal intelligence, often described in spiritual traditions as the Divine, the Source, Cosmic Consciousness, or God.

These three provinces of consciousness are, to varying degrees, present within all energy and mass throughout the universe. Recall that they are so labeled strictly for their specific functions. The singular difference among them is their energy level, with Cosmic Consciousness at an infinitely higher level than Immanent and Collective Consciousness.

This tripartite model provides a scaffolding upon which Transcendent Physics can be further built and scientifically explored. It integrates quantum mechanical insights while expanding beyond them, offering a more complete and consistent picture of reality.

The Role of Consciousness in Reality Formation

One of the most controversial yet compelling ideas in Transcendent Physics is that consciousness does not merely observe reality—it actively participates in its creation. This aligns with interpretations of quantum mechanics, such as the Von Neumann-Wigner hypothesis,[144] which posits that consciousness collapses the quantum wave function, thus bringing Werner Heisenberg's "potentia" into actuality.

Building on this concept, Transcendent Physics proposes a more direct relationship between consciousness and the structure of reality:

- Consciousness acts as the *primary force* that creates, organizes, and informs matter and energy in our four-dimensional universe.

- In highly complex entities such as the human body, it's highly likely that consciousness plays a role in managing, at least in part, the function of the multiplex systems created by its 33 trillion cells.

- Reality is *probabilistic rather than deterministic*, meaning that future events are shaped by Immanent Consciousness—that is, focused intention and attention, with relevant and necessary participation of Collective Consciousness and Cosmic Consciousness.

- The universe is inherently *teleological*, meaning that it has purpose and direction; and is largely driven by the evolution of consciousness to higher energy levels, the latter being the sole purpose of the universe, an eternal process that occurs via creation of universe after universe, ad infinitum, and driven by the eternal inextricable connection between physical and consciousness evolution.

This perspective challenges traditional materialist and naturalists viewpoints, suggesting that rather than being an epiphenomenon or a secondary effect of the brain, consciousness is fundamental—more fundamental than space-time itself. As Max Planck proclaimed nearly 100 years ago, "It is *intrinsic* to the universe."

The Mechanics of Transcendent Physics

To make Transcendent Physics fully operational, we must define key principles that govern its mechanics. Some of these principles have been previously discussed in this text. They extend quantum concepts into Transcendent Physics, while others introduce new postulates:

1. **The Primacy of Consciousness:** Consciousness is not an *emergent* by-product of matter and energy but an *intrinsic* property of existence, shaping reality at all levels by creation of all matter and energy throughout the universe.

2. **Nonlocality of Mind:** Just as quantum entanglement allows for instantaneous connections, consciousness is not bound by spatial and temporal constraints of the body and can influence distant events within the past, present, and future.

3. **Time as a Flexible Construct:** Past, present, and future are convenient constructs that allow us to function comfortably in our four-dimensional world. However, in True Reality, they exist simultaneously, layered upon each other in a higher-dimensional structure; therefore, consciousness has the potential to navigate this continuum and manifest its influence over all past and future space-time.

4. **Interdimensional Resonance:** Reality is structured in layers;[145] and consciousness can tune in to different dimensions through intention, attention, and consciousness frequency alignment and overlap.

5. **Manifestation Through Focused Intent:** Consciousness has the ability to collapse probability fields—wave functions—into specific outcomes, effectively bringing thought into form or form into thought.

These five principles lay the foundation for practical applications of Transcendent Physics, from bodily healing modalities to major technological advancements that harness nonlocal consciousness. They are most effectively practiced at higher energy levels of consciousness.

The Evolution of Consciousness: Humanity's Role

If Transcendent Physics is to be fully realized, humanity and, particularly, the scientific and philosophy communities, must undergo a much-needed evolution in the understanding, study, and use of consciousness. This requires a shift from a purely analytical approach to one that closely integrates the metaphysical conjugates of intuition, meditation, and expanded awareness with detailed mathematical and physical analyses—a path to solving the Hard Problem. This merging of physics, mathematics, and metaphysical philosophy will not be an easy transition, but it must begin immediately if humanity is to prosper from the immense potential of managed consciousness.

Scientific methodologies must evolve accordingly, embracing new tools for measuring and interacting with consciousness. Just as quantum mechanics required new experimental designs to validate its theories, Transcendent Physics will necessitate innovative approaches that extend beyond traditional instrumentation and current laboratory methods.

These efforts must be mindful that presently the most effective "sensor" and "choreographer" of consciousness is the consciousness of a human being, functioning at elevated energy levels via deep meditation. The implication is that the more successful transcendent scientists and philosophers will be those who embrace and practice deep meditation in conjunction with their research. This will provide efficient access to timeless wisdom from the confluence of specifically directed Immanent, Collective, and Cosmic Consciousness.

The integration of Transcendent Physics into human society could have profound implications:

- **Advancements in Health Care:** Understanding the consciousness-matter-energy relationship could lead to new healing modalities that harness and optimize mind-body-spirit interactions.
- **Breakthroughs in Energy Generation:** If consciousness can manipulate probability fields, it may be possible to extract usable clean energy from the zero-point energy (ZPE) field of the universe.[146] Richard Feynman and John Archibald Wheeler once calculated that the amount of energy that exists in the universe's zero-point field is so huge that it could provide clean, sustainable energy, essentially forever.[147,148,149]
- **Enhanced Human Potential:** Developing techniques to access higher states of consciousness could unlock extraordinary abilities such as elevated creativity, heightened intuition,

remote viewing, and nonlocal perception. The last two areas have been the subject of significant top-secret research efforts by the US and Russia and have demonstrated successful outcomes for psychic espionage.[150,151,152]

- **A New Ethical Framework:** Recognizing the interconnectedness of consciousness would encourage a shift toward greater cooperation, compassion, and responsibility on a planetary scale. Essentially, humanity would move closer to embracing the Six Cosmic Core Values of the universe.

The Future of Transcendent Physics

While quantum mechanics took decades to gain mainstream acceptance, the timeline for Transcendent Physics may be even longer. Societal inertia, skepticism, and the limitations of current scientific paradigms all pose significant obstacles. However, as history has shown, transformative ideas eventually find their way into collective understanding and acceptance.

A critical next step in the advancement of Transcendent Physics will be the development of experimental frameworks capable of testing and elaborating on its core principles.

Experimental Frameworks

This could include, for example:

- The development of means to normalize and standardize the human consciousness capability as a sensory means to quantitatively study consciousness.
- Advanced brain-wave and consciousness research, exploring the correlation between thought and physical reality, thereby more effectively connecting Transcendent Physics with neuroscience.
- Quantum experiments designed to test the influence of degree and quality of intention and attention on wave-function collapse.

- The exploration of nonlocal consciousness effects through controlled studies of telepathy, remote viewing, near-death experiences (NDEs), and out-of-body experiences (OBEs). Interviews with people who have been properly vetted to have had an NDE or OBE and are open to discussion and possibly hypnosis would be a fruitful area of research.[153] It has the potential to yield valuable insight concerning the nature of life, death, True Reality, and the purpose of all that exists within the universe—including the universe itself.

Research in Transcendent Physics will benefit from multidisciplinary efforts. This should include teams composed of theoretical and experimental physicists with philosophers of science whose training and experience are steeped in consciousness, physics, and metaphysics research.

By building upon the foundation laid by quantum physics, Transcendent Physics has the potential to revolutionize our understanding of reality and the universe. It may well be the next great scientific frontier—one that unifies the physical and the metaphysical into a coherent, actionable, "omniexistent" framework.

Conclusion

The transition from classical mechanics to quantum mechanics revolutionized physics, but it did not complete the journey. Transcendent Physics represents the next step—a framework that seeks to unify consciousness with the fundamental fabric of reality. Through this lens, the universe is not a cold, mechanistic void, but rather, a dynamic, intelligent system where consciousness plays a direct and participatory role. Paralleling James Lovelock's *Gaia Principle for planet Earth,*[154] we might conceive of applying and modeling this principle for the entire universe as a single functional "organism." As mentioned previously, scientists have tried to develop a single wave equation for the entire universe—why not a single equation for a single organism?

As humanity continues its journey of discovery, embracing this expanded perspective will not only enhance scientific understanding but also transform the way we experience ourselves, the world, the universe, and reality itself. The future of physics is not just about particles, energies, and forces; it's about its role in the very nature of being, perception, and the infinite potential that consciousness represents.

REFLECTION 24

What about God?

"Coincidence is God's way of remaining anonymous."

—Albert Einstein

In a world steeped in questions of existential significance, inquiry into the existence and nature of God endures as one of humanity's oldest, most contentious, and most profound pursuits. The religious frameworks ingrained in various cultures each provide their unique interpretations, constructing various narratives about the Divine.

However, the field of Transcendent Physics presents a secular philosophical approach for the nature of the Divine that transcends traditional views, offering a comprehensive understanding of physical and spiritual existence that seamlessly harmonizes with the essence of consciousness and the universe.

Transcendent Physics as a Scientific Philosophy

It's important to clarify from the outset that Transcendent Physics is not a religion. It's a philosophy grounded in scientific and metaphysic principles that elucidate the true nature of reality—perhaps, more to the point, the nature of True Reality. This philosophical approach respects various religious beliefs, acknowledging that many individuals find comfort in organized religion.

However, Transcendent Physics differs significantly from faith-based systems. It seeks to understand the underlying essence and function of consciousness that connects us all, offering an inclusive framework for inquiry rather than prescriptive doctrines and dogma on what one *must* believe and practice, lest there be negative repercussions here and hereafter. It focuses on consciousness because, as discussed throughout this text, it is the only *intrinsic* property of the universe—the source of all energy and matter. *It's not just an important thing; it's the only thing!* As a result, the answers are always *within*—not without.

Transcendent Physics invites those with divergent beliefs to explore the interconnectedness of all existence without the constraints often imposed by organized religions. While Transcendent Physics may challenge conventional doctrines, it does so not to undermine them but to expand the conversation around the Divine, offering a straightforward context aligned with the findings of modern science, philosophy, and the genesis and cosmic progression of the universe. To be sure, a significant amount of Transcendent Physics is currently based on Conjecturisms. However, with further research, this is projected to change.

The intent of this reflection is to offer a means to a clear and accurate understanding of the essence of the Divine, and through its access, illuminate a path to a more joyous and fulfilled life.

The Nature of the Divine in Transcendent Physics

Central to the philosophy of Transcendent Physics is the conception of Cosmic Consciousness. Far from depicting a deity that rewards or punishes based on moral behavior, Transcendent Physics views this consciousness as the ultimate source from which all forms of life derive. Cosmic Consciousness is understood as the foundational essence of reality itself—functioning as an omnipresent, omnipotent, and omniscient yet subtle force that permeates all existence—all beings, species, matter, and energy. Its essence embraces the Six Cosmic Core Values: truth, wisdom, understanding, forgiveness, love, and compassion.

This notion aligns with the Panpsychist view, which posits that consciousness is ubiquitous and present in all things—both sentient and insentient. Instead of perceiving the Divine as a distant being residing in a realm separate from our own, Transcendent Physics encourages an understanding of the Divine as interwoven throughout the entire fabric of the universe, participating intimately in the dance of all creation and all functioning existence. This view helps to dissolve the duality that often separates the sacred from the mundane, thereby emphasizing their profound unity—a necessity for true physical, emotional, and spiritual growth—and eventually the achievement of what some might call *enlightenment*.

Karma and Cosmic Consciousness

The principles of karma also feature prominently in the discourse around the Divine as viewed within Transcendent Physics. Karma is not viewed as a system of reward and punishment, but rather a natural law—an inevitable consequence of one's free-will actions that shape reality. Proceeding in a kind of quantum mode of action, each deed, thought, and intention generates cosmic ripples or wavelets throughout the fabric of consciousness, influencing not only the individual but the greater collective as well.

These wavelet are eventually reflected back by Collective Consciousness, returning to their source, providing input on those actions that were and were not based on the Six Cosmic Core Values. This iterative process engenders the path for a positive evolution of consciousness to higher energy states.

The errors and misunderstandings often associated with the concept of karma arise when one perceives it through a dualistic lens. Instead of seeing it as punitive, Transcendent Physics encourages viewing it as a resonant echo of the interconnectedness of all beings—a reflection of how our choices shape reality collaboratively—hence, the role of Collective Consciousness. Just as the individual is inseparable from the whole, so, too, are our actions intertwined with the grand tapestry of all existence.

In this light, prayer and meditation assume a transformative role. When individuals engage in sincere prayer—viewed as a form of meditative practice—they can align their intentions with both Immanent Consciousness (consciousness inherent in all matter and energy) and *Cosmic Consciousness* (universal intelligence). This alignment activates pathways for manifestation that affirm the principles of karma and mutual reciprocity throughout the universe.

Connections Between Science and Spirituality

Transcendent Physics seeks to bridge the perceived chasm between science and spirituality, revealing a comprehensive framework that honors both within any inquiry. The marvels of quantum physics have already begun to challenge classical materialism, suggesting that consciousness is not merely an *emergent* by-product of neurological brain functions but rather an *intrinsic* aspect of the universe, creating and influencing all True Reality.

As we delve deeper into Transcendental Physics' philosophical architecture, we can further elucidate concepts such as quantum entanglement and the Observer Effect. These phenomena suggest that consciousness plays a crucial role in shaping physical reality, establishing a feedback loop whereby observation influences the outcomes that occur, even over cosmic distances. This acknowledgment—that consciousness participates in creating reality—parallels traditional spiritual notions that ascribe creative power to earnest intention and attention. However, in Transcendent Physics, this power is considered to actuate in a collaborative action with the Divine—that is, Cosmic Consciousness.

Belief systems that maintain a dichotomy between the material and the spiritual often fall short of capturing the intricate interplay at work, where the Divine and scientific coexist as layers of a unified experience. By embracing Transcendent Physics, individuals may find an integrated understanding that honors both their spiritual inclinations and their scientific curiosity.

A New Ethical Framework

Emerging from Transcendent Physics is a new ethical framework rooted in the understanding of universal interconnectedness through the fundamentals of consciousness. If we accept that all beings are manifestations of a singular undifferentiated consciousness, our approach to morality undergoes a transformation. Coherence with the Six Cosmic Core Values becomes an imperative to foster harmony and progress within the universe.

In this new schema, ethical behavior is not dictated by rigid laws, decrees, or dogma, which most often were created at the hands of men—sometimes with the best of intentions, and sometimes not. Rather, this framework unfolds naturally as individuals align with their inherent consciousness and joint purpose to support the Purpose of the Universe—namely, *continuously increasing its total consciousness energy level.* Compassion and empathy become natural responses, incentivized by a desired natural functioning as a *Whole,* rather than as *One*. In our nondualistic universe, all of this can come quite naturally from within, not externally.

So, Then . . . What about God?

Through the lens of Transcendent Physics, the question of God need not remain shrouded in ambiguity. Hopefully, by what has transpired throughout the Reflections in this book, what many traditions believe as God, Transcendent Physics discerns as Cosmic Consciousness—a vast, omnipresent, *omni-intelligent* awareness that both transcends and permeates the entire universe—including you.

Was There a Beginning?

In Reflection 9, Conjecturism 1, Transcendent Physics posits that there was no beginning to the eternal formation of universes. There is another possibility that would not significantly change the details of cosmological physics, but recognize the eternal and infinite presence of Cosmic Consciousness as God throughout the universe.

What if, in the beginning, there actually was a beginning, and all that existed in an ethereal realm of "nothingness" was Cosmic Consciousness—God, if you wish. And God "decided," or perhaps a unique, negative entropy transition occurred that formed a singularity, a "primeval atom" containing all of the Cosmic Consciousness, its eventual explosive expansion, and formation of the first universe.

Recall that consciousness is *intrinsic* to the universe and the source of all its energy and matter. Thus, the universe would begin its formation by means of the Big Bang and its further growth by the inextricable interconnected evolution of physicality and consciousness, based intimately on the Six Cosmic Core Values, as previously described.

The remaining processes of "creation"—formation, expansion, contraction, dissolution, and formation of the next universe—would occur as previously described . . . evolving eternally with physical and consciousness evolution. This Conjecture might resonate with you if you have a *creationist* predilection.

Not a Religion

While this emerging science honors the historical and cultural significance of organized religion for those who find meaning in it, as mentioned previously, it's important to recognize that Transcendent Physics is not a religion. There are no rules, dogma, and laws that *must* be followed. On the contrary, deep thought, analysis, and debate of the stated Conjecturisms for Transcendent Physics are encouraged and important for its further refinement. The continued evolution of this science is based on deep universal thought. It is a cosmic philosophy grounded in science yet reaching beyond its current limits—a philosophy and metaphysical perspective that integrates physical reality with consciousness as the primary substrate of all existence.

Cosmic Consciousness exists simultaneously both beyond space-time and within it. It is both *Transcendent*—existing beyond all dimensions—and *Immanent*—alive within every atom, all energy, every breath, every thought, and present in every being. This nondualistic perspective reveals a universe in which all things are interconnected by a conscious energy field that expresses itself through the Six Cosmic Core Values.

This interconnection is not metaphorical. It's energetic, intrinsic, and represents True Reality. It holds transformative potential for each of us. To access it, we must quiet the turbulence of fear, desire, and distraction that defines so much of our experience in our four-dimensional world. Through focused intention, disciplined attention, choiceless awareness, meditation, and aligned action, we can consciously attune ourselves to this deeper reality and create the fulfillment we seek.

Those who currently meditate on, or pray to, a specific spiritual avatar, such as Jesus, Allah, Mary, Buddha, Vishnu, or Adonai, may feel some discomfort directing their spiritual practice toward Cosmic Consciousness. This is primarily because most of us with a religious belief system have been programmed since birth with a preferred name for the primary spiritual entity at the apex of our religion. In a practice within Transcendent Physics, you might consider, for example, Cosmic Christus, Cosmic Allah, Cosmic Buddha, Cosmic Vishu, Cosmic Adonai, or any other name that honors your respect for a higher power in concert with the philosophy of Transcendent Physics.

A Closing Reflection

As we conclude our exploration of consciousness, consider this: Anything is possible. Whether it's discovering your Life Purpose, deepening your understanding of consciousness, reality, and the universe, or manifesting your most heartfelt desires—the potential lies within you.

The only condition is to align your journey with the Six Cosmic Core Values and the Two Laws of Manifestation. Let your *intention* serve as the fuel that energizes the field of *consciousness*. Let your *attention* become the guiding force that shapes your reality.

As part of your meditation and Choiceless Awareness activities, consider offering this thought: "I place my intentions into the vast Ocean of Consciousness and its infinite possibilities. I allow the universe to move through me, knowing that my actions and desires are supported by the infinite intelligence of Cosmic Consciousness (God)."

That's the invitation of Transcendent Physics: to *awaken within*. Cosmic Consciousness is not just "out there." It's also within . . . *you*! And always has been.

EPILOGUE

I hope that the Reflections on Transcendent Physics expressed in this book have presented you with some important ideas and concepts, and at a minimum, stimulated your thoughts about this emerging field of science and philosophy. If so, I trust that you will intimately connect with the following thoughts from Deepak Chopra:[155]

Reality

"What is reality? Is it what we can experience with our senses? Or is it something else, something deeper? Consider for a moment, . . . 'Where was your newborn before she came into this world?' Just because we cannot see or touch something does not mean it doesn't exist. It simply means we can't experience it in our four-dimensional world at this moment.

"The same holds true for whatever you consider abundance—limitless love, unbounded joy, optimal health, or greater material possessions. Just because you're not experiencing what you desire in this moment, doesn't mean it can't exist for you.

"We live our lives primarily on the physical plane, which includes everything that has form or substance—what we commonly refer to as matter. Some look at this physical realm and see lack. Others, see abundance. Some feel abundant. Others feel limited based on certain messages they may have received in the past.

"While we believe matter to be reality, we often view reality as changeless. Yet, matter and form continually change. So, what then is reality?

"As we exist in the physical plane, we also live at the level of the mind. The mind is that part of us that perceives, thinks, reasons, and evaluates. The mind is potential energy, and because of this we can change our physical world simply by changing our thoughts and beliefs.

"Beyond the mind at the deepest level of consciousness, resides the spirit. This is the part of us that is eternal, unchanging, and imbued with pure unlimited potential. Tapping into this potential is what enables us to manifest miracles.

"The world 'out there' may appear to be objective, but in fact, it is subjective, a construct of our own interpretations. Our words, thoughts, and beliefs create specific results in the physical world. Therefore, abundance does not spring from an external source; rather, it comes from within—from thought, intention, attention, and expectation.

"You, as a powerful creative being, can change your perceptions, change your mind, and change your reality by transforming your thoughts and expectations. You can begin to speak and understand nature's language and harness its infinite flow of energy, information, and intelligence to enjoy the great bounty of spirit that is your birthright."

Enjoy your journey; make a difference.

James A. Cusumano
Prague

APPENDIX I

Electron Double-Slit Experimental Data

The exposure time for each photograph increases from (a) to (e).

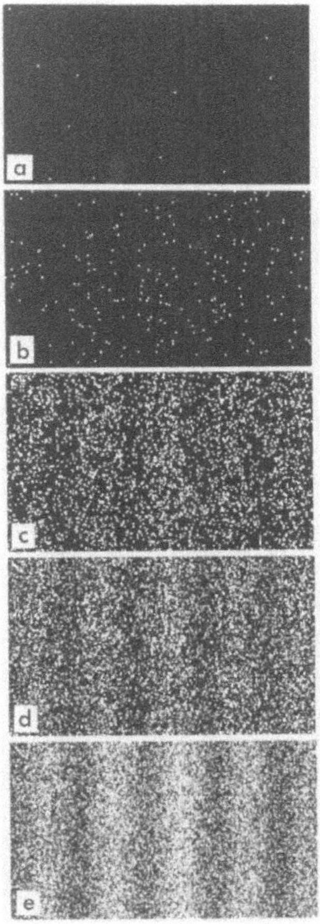

APPENDIX II

Love: The Most Powerful Force in the Universe

"Love is a liberating force.
Attachment is an entangling process."

—Isha Sadhguru[156]

Throughout history, the greatest minds and wisdom traditions have asserted that love is the fundamental force that governs many if not all actions in the universe. But how can this be? How can love extend beyond human emotions into the very fabric of cosmic existence?

What Is Love?

Love is notoriously difficult to define. The dictionary offers several meanings: "a strong feeling of affection; a great interest and pleasure in something; or a deep feeling of sexual love for someone." However, these descriptions fail to capture its multidimensional essence and universal presence.

Love is more than a sentiment, an emotion, or even a biological drive for reproduction. It is a pragmatic and necessary force because, as Transcendent Physics teaches us, all things—living and nonliving—possess dimensions of consciousness and are universally interconnected and inextricably interdependent, now and forever.

Love serves as a catalytic force, enabling interactions that benefit the *Whole*, guiding an infinite evolutionary journey toward *Oneness*. Just like a catalyst in a chemical reaction, love lowers and often eliminates the barriers to transformation, creating something of immense value while avoiding negative consequences. In the science of catalysis, that's known as maximum *specificity* and *selectivity*. The greatest achievements of humankind have always involved an element of love at their core.

The Language of Love

One of the challenges in understanding love is that it manifests in countless ways, yet English uses but a single word to encapsulate them all. In contrast, Sanskrit has 96 words for *love*, Farsi has 80, and Greek has at least three—*eros* (romantic love), *philia* (deep friendship), and *agape* (universal, selfless love). The English language's limitations often lead to confusion, making it difficult to communicate the depth of what is meant in a given situation.

Love can describe a range of experiences, from fleeting pleasures: "I loved that meal," to deep emotional commitments: "I love her with my whole heart and soul." However, because love is an attractive force of consciousness energy, it extends even further, beyond human relationships and emotions, into the fundamental interactive forces of physics and the biological sciences.

Love in the Fabric of the Universe

The attraction within a hydrogen atom between its negatively charged electron whizzing around the atom's nucleus and a positively charged proton resting within the nucleus can be seen as a primal form of love. Without this fundamental attractive interaction, atoms would not form, and neither would the universe as we know it. Similarly, magnetism, which binds the North and South Poles, mirrors the dance of attraction and unity. A similar primal form of love can be envisioned for the universal force of gravity.

Love also manifests biologically at a rudimentary level. When certain cells in the body sense that one of their neighbors is in distress, they share essential nutrients, ensuring the survival of the whole organism. This process mirrors the fundamental principles of James Lovelock's Gaia hypothesis, where Earth's biosphere behaves as a single organism and self-regulates to sustain life. It has done so endlessly over billions of years, even after five major mass extinctions of life on the planet. Its *life force* cannot be destroyed, driven by the cosmic love-attraction power to renew, grow, and survive.

From a philosophical standpoint, love represents compassion, kindness, and the selfless desire for the well-being of others—similar to the Six Cosmic Core Values of the universe. It's the glue that binds communities, the force that inspires acts of great courage, and the energy that fuels human creativity and connection.

Love in Human Endeavors

Love is a powerful force not only in personal relationships but also in business, leadership, and social structures. In the business world, love translates to trust, commitment, and genuine care for others. Great leaders inspire because they care deeply about their people, and the converse being true, even when their leaders are demanding. True innovation, creativity, and lasting success are built on this foundation of trust and shared purpose.

The Role of Selfishness

If selfless love drives the evolution of consciousness energy, what about selfishness? Even selfishness is a form of love—love of self, or more precisely, love of the ego. This form of love often leads to destructive behaviors, conflicts, and suffering.

However, there is a distinction between destructive selfishness and compassionate self-interest. When one acts in service to others, not only do they help one another, but they also nourish their own soul. The Wisdom Seekers of the East note that *you must learn to love yourself before you can love someone else*—meaning that respect and gratefulness for who and what you are precedes your ability to give to others. The Prayer of Saint Francis, beautifully states: "For it is in giving that we receive."[157]

The Evolutionary Role of Love

The universe's evolution is a continuous interplay of consciousness, connection, and the search for unity. The great mission of existence is the evolution of consciousness toward higher energies of Oneness, and love is the catalytic force field that facilitates this process. Every act of love—whether between atoms, people, or ecosystems—propels this grand evolutionary journey forward.

Ultimately, love is the underlying principle of existence itself. The more we recognize its universal presence, the more we align ourselves with the deepest truths of reality and the Six Cosmic Core Values.

Love is not just an emotion—it is consciousness in action, a form that is the essence of all things, the driving force behind the cosmos, and the ultimate path to enlightenment.

Acknowledgments

My deepest admiration and gratitude to those special souls who have graced my life path and made a positive difference by showing me how I might do the same for others.

Suggested Reading

Sean Carroll, *Something Deeply Hidden: Quantum Worlds and the Emergence of Spacetime*, Dutton, New York, Sept. 1, 2020.

D. J. Chalmers, Facing Up to the Problem of Consciousness, *Journal of Consciousness Studies*, 2(3), 200–19, 1995.

Federico Faggin, *Irreducible: Consciousness, Life, Computers, and Human Nature*, Essentia Books, June 1, 2024.

Philip Goff, *Galileo's Error: Foundations for a New Science of Consciousness*, Vintage Press, London, Aug. 6, 2020.

Philip Goff, *Why? The Purpose of the Universe*, Oxford University Press, Nov. 9, 2023.

Bernard Haisch, *The God Theory: Universes. Zero-Point Fields, and What's Behind It All*, Weiser Books, April 1, 2009.

Annaka Harris, *Conscious: A Brief Guide to the Fundamental Mystery of the Mind*, Harper, New York, June 4, 2019.

Donald Hoffman, *The Case Against Reality: Why Evolution Hid the Truth from Our Eyes*, W. W. Norton & Company, New York, Nov. 8, 2022.

Sabine Hossenfelder, *Existential Physics: A Scientist's Guide to Life's Biggest Questions*, Atlantic Books, London, UK, July 6, 2023.

Benardo Kastrup, *Why Materialism Is Baloney: How True Skeptics Know There Is No Death and Fathom Answers to life, the Universe, and Everything*, Iff Books, New Alresford, Hampshire, UK, April 25, 2014.

Thomas Nagel, *What Is It Like to Be a Bat?* 50th anniversary edition, Oxford University Press, London, June 4, 2024.

Bruce Rosenblum, Fred Kuttnes, and Henry Stapp, *Quantum Physics of Consciousness: The Quantum Physics of the Mind Explained*, Science Publishers, Feb. 16, 2017.

David Skrbina, *Panpsychism in the West, revised edition*, The MIT Press, Sept. 22, 2017.

Henry P. Stapp, *Mindful Universe: Quantum Mechanics and the Participating Observer (The Frontiers Collection)*, 1st Edition, June 13, 2007.

Galen Strawson, *Things That Bother Me: Death, Freedom, the Self, Etc.*, *The New York Review of Books*, March 13, 2018.

Endnotes

[1] *The Observer*, Interview with Max Planck on the Nature of Quantum Mechanics and Consciousness, Jan. 25, 1931.

[2] These included Erwin Schrödinger, John Von Neumann, Eugene Wigner, and Werner Heisenberg, among others.

[3] Philip Goff, *Galileo's Error: Foundations for a New Science of Consciousness,* Vintage Press, London, Aug. 6, 2020.

[4] Thomas Nagel, *What Is It Like to Be a Bat?* 50th anniversary edition, Oxford University Press, London, June 4, 2024.

[5] The principles of Transcendent Physics are summarized in Reflections 8 and 9.

[6] Nisargadatta Maharaj, *I Am That,* Chetana Pvt. Ltd., Mumbai, India, Aug. 1, 1999.

[7] Patanjali, *How to Know God: The Yoga Aphorisms of Patanjali*—translated with commentary by Swami Prabhavananda and Christopher Isherwood, Vedanta Press, Hollywood, CA, Aug. 30, 1953.

[8] Abdullah Yusuf Ali, *The Holy Qur'an*, Wordsworth Editions Ltd., Ware, Hertfordshire, England, April 5, 2001.

[9] *The Torah in English,* ISBN-13: 979-8398759907, independently published June 19, 2023.

[10] James A. Cusumano, *I Can See Clearly: Rise of a Supernatural Hero—Book 1 of the Luc Ponti Chronicles,* Waterside Productions, Cardiff-by-the-Sea, CA, 2020.

[11] James A. Cusumano, *Seagull's Revenge: Beyond Fear—Book 2 of the Luc Ponti Chronicles,* Waterside Productions, Cardiff-by-the-Sea, CA, 2021.

[12] James A. Cusumano, *Cosmic Contact: The Next Earth, a New Beginning—Book 3 of the Luc Ponti Chronicles,* Waterside Productions, Cardiff-by-the-Sea, CA, 2023.

[13] Dan Brown, *The Secret of Secrets,* Doubleday, New York, 2025. Film rights have been signed with Netflix for an eight-episode series.

[14] James A. Cusumano, *Dance to the Rhythm of the Cosmos: Tapping into the Infinite Power of Consciousness,* Waterside Productions, Cardiff-by-the-Sea, CA, 2024.

[15] Ibid.

[16] One could make the argument that *consciousness* would be a better descriptor than *conscious.* However, if you dwell on the difference, you'll see that a subtle but more fundamental meaning is intended.

[17] Deepak Chopra, The Riddle of a Conscious Universe, *Medium*, Feb. 26, 2024.

[18] Singularity—a point at which a mathematical function has an infinite value, especially in space-time when matter is infinitely dense, such as at the center of a black hole. In 1927, Belgian astronomer Georges Lemaître extrapolated Einstein's General Relativity

equations to time zero and posited that the universe was formed by expansion of a singularity, which he called a "primeval atom" or a "Cosmic Egg." The residual heat from the Big Bang, which stretched and cooled over 13.8 billion years by cosmic expansion, was projected to leave an afterglow of approximately 2.7 Kelvin, or about -270°C, emitting blackbody thermal radiation in the microwave portion of the electromagnetic spectrum. In 1964, this afterglow was discovered by American radio astronomers Arno Penzias and Robert Wilson. It was found to pervade all of space and is known as the Cosmic Background Radiation (CBR). For their discovery, they were awarded the Nobel Prize in 1978.

[19] Op. cit., *The Observer* interview with Max Planck.

[20] It's quite telling that the term *metaphysics* means "beyond physics." However, in the interest of broader acceptance, and to circumvent unnecessary confrontation, we avoid that term in favor of *Transcendent Physics.*

[21] Without getting into mathematical complexities, we can say that the degree and strength of this overlap can be calculated precisely by evaluating what in quantum mechanics is called an *overlap integral.* The value of this integral—namely, the quantitative value of the degree of overlap of Immanent Consciousness with Collective and Cosmic Consciousness determines the strength of the interconnection, and therefore, the ease of connection and interaction.

[22] "The Mistress of Vision" poem by Francis Thompson, no. XXI.

[23] The word *Divine* is not intended to connote a connection with any organized religion, but to highly elevated levels of consciousness with capabilities well beyond those demonstrated by humanity and therefore, often referred to as *God.*

[24] To be precise, since energy and mass are formed from consciousness, the thing that is universally and eternally constant is the sum of existing consciousness plus the consciousness equivalent to that used to form energy and mass.

[25] The *Big Bounce* immediately follows the *Big Crunch*, leading to formation of the next universe.

[26] Our Best Understanding of the Universe May be Wrong, *New Scientist*, March 22, 2025, p. 8. This research suggests that the cosmological constant (λ) is *not* constant. It is decreasing so that the rate of acceleration of the universe's expansion is slowing down because Dark Energy, which *is* the cosmological constant (λ) and controls the expansion, is decreasing in its strength. As this process continues, the universe expansion will stop and change to a contraction, eventually creating the Big Crunch and forming the next singularity, which then transitions to the Big Bounce expanding to form the next universe.

[27] Martin Bojwald, Big Bang or Big Bounce?: New Theory on the Universe's Birth, *Scientific American*, Oct. 1, 2008.

[28] David Bohm, *Wholeness and the Implicate Order*, Routledge & Kegan Paul, New York, 1980.

[29] Quantum number—a number that occurs in a quantum mechanical expression for the value of some quantized property of an atomic or subatomic particle—for example, an atom or molecule, and can only have certain integral or half-integral values.

[30] Julien Offray de la Mettrie, *L'Homme Machine,* Open Court Publishing Co., Chicago, 1912, first published in 1747.

[31] Walt Whitman, "Song of Myself," included in *Leaves of Grass*, 1855.

[32] Kurt Ebert, How Quantum Physics Looks at the Energetic Mind, *Attractor Field Therapy*, June 7, 2019.

[33] D. J. Chalmers, Facing Up to the Problem of Consciousness, *Journal of Consciousness Studies*, 2(3), 200–19, 1995.

[34] Martin Myles Filler, The Late Show, *The New York Review of Books*, June 11, 2009.

[35] Blackbody Radiation consists of photons of light given off by an object based on its temperature. It is electromagnetic radiation emitted by a perfect absorber and emitter of electromagnetic radiation, directly related to the temperature of the object.

[36] Tom Clancy, Goodreads.

[37] The Quantum Thomist, Dec. 31, 2017.

[38] In quantum mechanics, this is referred to as the collapse of the wave function to the highest-probability state of being.

[39] Stephen Hawking estate, https://wwwhawking.org.uk/.

[40] As discussed in Part II, physical events and/or objects may be expressed by a complex mathematical *equation* or *function*—these two words are synonymous—called the quantum wave equation or quantum wave function. The probability of a certain event occurring or materializing a certain object is equal to the square of the wave function describing the event. If Ψ represents the wave function for the event, then if you solve for Ψ^2, you can calculate the precise probability of you observing that event.

[41] As discussed in Part II, your soul is not a theological or religious entity. It is your Immanent Consciousness, your True Reality, which provides your awareness and is connected to all others through Collective Consciousness. So, you could say it's a *spiritual* entity.

[42] *Choiceless Awareness:* This exercise was popularized by Indian philosopher Jiddu Krishnamurti.

[43] Op. cit., Nisargadatta Maharaj, pp. 288–89.

[44] Technical aspects of Transcendent Physics are summarized in Reflections 22 and 23.

[45] Walt Whitman, in *Leaves of Grass—A Glimpse of the Soul*, 1855.

[46] *Inextricability* of physical and consciousness evolution is a fundamentally profound concept. The bottom line is that without physical evolution, there could be no consciousness evolution, and conversely—which means that without either, there would be *nothing*.

[47] Ibid.

[48] The word *Conjecturism* is meant to differentiate it from a simple conjecture, which is a quickly expressed thought or conclusion. In contrast, a Conjecturism is a well-thought-out, intelligently researched, and extensively analyzed conjecture. However, to be sure, it is not a proven fact. Conjecturisms are generally found in mathematics, philosophy,

logic, and the sciences to describe the meaning or essence of challenging concepts. For example, after more than 350 years, Fermat's Conjecture (1637)—actually a Conjecturism—was finally proven to be correct by mathematician Andrew Wiles of Oxford University. Now it's called Fermat's Theorem. These 10 Conjecturisms are based on decades of my multidisciplinary research in quantum physics, philosophy, metaphysics, cosmology, and Eastern Wisdom thought and philosophy. Each Conjecturism was assessed numerous times in conjunction with deep meditation at high levels of consciousness.

[49] These 10 Conjecturisms are a distillation of key elements from Chapter 1 of my book *Dance to the Rhythm of the Cosmos,* Waterside Productions, Cardiff-by-the-Sea, CA, 2024.

[50] Paulo Coelho, *The Alchemist—Deluxe Edition*, HarperOne, New York, April 15, 2025.

[51] *Transubstantiation* is a mystical, unexplainable change in form, substance, or appearance of one substance into another. Medieval alchemists attempted the *Transubstantiation* of lead into gold.

[52] In an interview with *The Observer* on January 25, 1931, Max Planck, often considered the "father" of quantum mechanics said, "I regard consciousness as fundamental. Matter and energy are derived from consciousness. We cannot go beyond consciousness. Everything we talk about, everything that exists, postulates consciousness." A number of Planck's colleagues, all astute and accomplished quantum physicists, expressed similar views. These included Erwin Schrödinger, John Von Neumann, Eugene Wigner, and Werner Heisenberg, among others.

[53] David Skrbina, *Panpsychism in the West,* revised edition, The MIT Press, Sept. 22, 2017.

[54] Masaru Emoto, *The Hidden Messages in Water*, Atria Books, New York, 2004. Most Panpsychists posit that the consciousness in nonliving entities is primitive. Emoto's research on consciousness associated with water and ice crystals suggests that some entities—for example, water, have more highly developed consciousness than one might expect.

[55] James P. Kowall, How Consciousness Creates Life, Medium.Com, May 4, 2025. Recent work by Kowall, a theoretical physicist and physician, provides a theory of consciousness that has some similarities to Trans-Immanent Panpsychism.

[56] During the 1960s, John Wheeler concluded that space-time was not an intrinsic property of the universe. He was convinced that it was an emergent entity—but from "what" did it emerge? This inspired numerous research projects to find that "what," but to date, all have been unsuccessful. Through the lens of Transcendent Physics, it's clear that the "what" is pure consciousness.

[57] The late Jesuit theologian and philosopher Teilhard de Chardin is said to have functioned with the Three Faces of God. The *First-Person Face* is the subjective, mystical, inner face of God, which we may call our True Self (our *Soul or Immanent Consciousness*). To the degree that we lose our identification with our egoic personality, we instead identify with our Divine-Self (*Cosmic Consciousness*).

[58] Joran Slane Oppelt, Pantheism or Panentheism?, *Progressive Christianity*. March 12, 2019.

[59] Pantheism is the identification of God with creation in such a way that the two are indistinguishable. Panentheism means that God is present in all creation by virtue of His omnipresence and omnipotence, sustaining every creature in being without being identified with any creature.

[60] Op. cit., Teilhard de Chardin and the *Three Faces of God.*

[61] Recall the words of Jesus Christ, who while speaking to a multitude, was asked, "How is it that you can perform miracles?" His humble response was, "All that I do, you can do, and more" (John 14:12).

[62] Our Best Understanding of the Universe May be Wrong, *New Scientist*, March 22, 2025, p. 8. This research suggests that the cosmological constant (λ) is *not* constant. It is decreasing so that the rate of acceleration of the universe's expansion is slowing down because Dark Energy, which *is* the cosmological constant (λ) and controls the expansion, is decreasing in its strength. As this continues, the universe expansion can stop and change to a contraction, and eventually creates the Big Crunch, followed by the Big Bounce.

[63] Frankline Misango Oyolo, Theoretical Astrophysics Research on How The Big Crunch Builds to the Big Bounce Theory, *International Collegiate Journal of Science*, October 4, 2022.

[64] For a comprehensive study and analysis supporting the existence of reincarnation, see the book (especially Chapters 1 and 2) by Leslie Kean, *Surviving Death: A Journalist Investigates Evidence for an Afterlife*, Crown Publishers, New York, March 6, 2018.

[65] Piotr Uklanski, *Interview Magazine,* March 24, 2011.

[66] Annaka Harris, *Conscious: A Brief Guide to the Fundamental Mystery of the Mind,* Harper, New York, June 4, 2019.

[67] Philip Goff, *Why? The Purpose of the Universe,* Oxford University Press, Nov. 9, 2023.

[68] Op. cit., Philip Goff, *Galileo's Error.*

[69] Similar to Panpsychism, but additionally exists beyond space-time, meaning the Universal Consciousness energy field is additionally posited to exist beyond the realm of the physical universe.

[70] Op. cit., David Skirba.

[71] Eben Alexander, *A Neurosurgeon's Journey into the Afterlife*, Simon & Schuster, New York, Oct. 23, 2012.

[72] Leslie Kean, *Surviving Death, A Journalist's Investigates Evidence for an Afterlife,* Crown Publishers, New York, March 6, 2018.

[73] Axel Balthazar, *Project Stargate and Remote Viewing Technology: The CIA's Files on Psychic Spying*, Adventures Unlimited Press, Aug. 26, 2020.

[74] Stargate Project (US Army Unit), Wikipedia.

[75] Deepak Chopra, *Quantum Healing: Exploring the Frontiers of Mind/Body Medicine,* revised, Bantam, New York, Nov. 17, 2015.

[76] Deepak Chopra and Rudolph E. Tanzi, *The Healing Self: A Revolutionary New Plan to Supercharge Your Immunity and Stay Well for Life: A Longevity Book*, Harmony Publishers, Harmony, June 16, 2020.

[77] Andrew Weil, *Spontaneous Healing: How to Discover and Enhance Your Body's Natural Ability to Maintain and Heal Itself*, Random House, New York, April 23, 1996.

[78] William Cromie, Harvard Gazette, *Meditation Changes Temperature*, April 18, 2002.

[79] Curt Newton, Meditation and the Brain, *MIT Review*, Feb. 1, 2004.

[80] Michelle Carr, Engineering Our Dreams, *Scientific American*, Jan. 2025, p. 46.

[81] Op. cit., Deepak Chopra.

[82] Tania Stoltz, Ulrich Weger, and Marcelo da Veiga, Consciousness and Education: Contributions by Piaget, Vygotsky and Steiner, *Frontiers in Psychology*, Vol. 15, Aug. 5, 2024.

[83] Sally Gillespie, *Climate Crisis and Consciousness: Re-imagining Our World and Ourselves*, Routledge, Oct. 28, 2019.

[84] David T. Risser, *Collective Moral Responsibility*, Internet Encyclopedia of Philosophy.

[85] Walter Glannon, *The Ethics of Consciousness (Elements in Bioethics and Neuroethics)*, Cambridge University Press, July 14, 2022.

[86] Deepak Chopra, *The Seven Spiritual Laws of Success: A Pocketbook Guide to Fulfilling Your Dreams (One Hour of Wisdom)*, Amber-Allen Publishing, Jan. 10, 2015.

[87] Albert Einstein's comment concerning the Observer Effect, in a letter to Abraham Pais, July 1949.

[88] Bruce Rosenblum and Fred Kutner, The Observer in the Quantum Experiment, https://arxiv.org/pdf/quant-ph/0011086.

[89] V. Jacques, E. Wu, F. Grosshans, F. Treussart, P. Granier, A. Aspect, and J.F. Roch, Experimental Realization of Wheeler's Delayed-choice Gedanken Experiment, *Science*, 315(5814):966, 2007.

[90] To be precise, spin is a quantum property of subatomic particles such as electrons and is most often represented by a quantity known as the *spin quantum number*. Although useful for modeling, it would be incorrect to think of electrons as little balls spinning either to the left or to the right. Spin is simply a nominal intrinsic property of electrons and gives rise to a host of important and often practical phenomena such as magnetism and nuclear magnetic resonance (NMR), to name just two.

[91] Sabine Hossenfelder, The Wave-function of the Universe Might Finally Be Calculable, YouTube, Feb. 6, 2025.

[92] Bruce Rosenblum and Fred Kutter, *Quantum Enigma: Physics Encounters Consciousness*, Oxford University Press, Oxford, England, 2006.

[93] Richard Feynman, Physics Forums.

[94] A quote from a 1954 letter by Albert Einstein to his friend Michael Besso. It highlights Einstein's profound and ongoing struggle with the nature of photons.

[95] EurekaAlert! *Feynman's Double-Slit Experiment Brought to Life,* March 13, 2013.

[96] *Science News,* Can we detect quantum behavior in viruses?, March 11, 2010.

[97] Copenhagen Interpretation of Quantum Mechanics, *Stanford Encyclopedia of Philosophy,* May 31, 2024.

[98] Many-Worlds Interpretation of Quantum Mechanics, *Stanford Encyclopedia of Philosophy,* Aug. 5, 2021.

[99] Op. cit., David Bohm.

[100] Stuart Hameroff, Quantum computation in brain microtubules? The Penrose–Hameroff 'Orch OR' model of consciousness, *Philosophical Transactions of the Royal Society A*, Aug. 15, 1998.

[101] Op. cit., David Bohm.

[102] Nucleosynthesis is the creation of new atomic nuclei, the centers of atoms that are made up of protons and neutrons. Nucleosynthesis is a nuclear fusion process and first occurred within a few minutes after the Big Bang. At that time, a quark-gluon plasma, a soup of particles known as quarks and gluons, condensed into protons and neutrons. Later on in the early formation of the universe, the remainder of the 92 naturally occurring elements of the periodic table were formed in the same fusion process within massive plasma stars and quasars. Subsequent galactic supernovae distributed these elements throughout the universe.

[103] The Four Fundamental Forces of Nature: The electromagnetic field, the strong nuclear force, gravity, and the weak nuclear force.

[104] Neil Bohr's work is nicely described in his book *The Theory of Spectra and Atomic Constitution; Three Essays,* pp. 1–18, Cambridge University Press, 1922.

[105] Atkins, P., The Rich Legacy of Erwin Schrödinger. *Nature* 326, 913–14 (1987).

[106] Throughout Part II, you will see that certain phenomena, such as Bohr's atomic theory, appear multiple times. This is by design, as often for those seeing quantum concepts for the first time, it helps to see them repeated in different formats in order to appreciate their implications for reality.

[107] Although Schrödinger struggled most of his professional life with the metaphysical aspects of quantum mechanics, he mellowed in later years as one can see in his famous book, *What Is Life*, which contains a number of philosophical and metaphysical ideas.

[108] De Broglie hypothesized that since an electron shows both *wave* and *particle* behavior, the circumference of an electron's circular orbit ($2\pi r$, r is the radius of the circular path) should be an integral multiple(n) of wavelengths(λ). Therefore, $2\pi r = n\lambda$. De Broglie determined that the wavelength is given by $\lambda = h/p$, where the momentum (p) is $p = mv$ and therefore $\lambda = h/mv$, where m is the mass and v is the velocity of the electron. By substitution, $2\pi r = nh/mv$ or, $mvr = nh/2\pi$.

[109] Fritjof Capra, *The Tao of Physics: An Exploration of the Parallels between Modern Physics and Eastern Mysticism*, Shambala Publishing, Boulder, CO, Aug. 13, 1991.

[110] As it turns out, in modern quantum theory, it is not possible to specify exactly both the electron orbit and its angular momentum simultaneously, as Bohr did in

developing his model. This is a violation of Heisenberg's Uncertainty Principle, which had yet to be discovered at the time Bohr developed his model.

[111] Angular momentum is a measure of an object's rotational motion, similar to how linear momentum describes motion in a straight line.

[112] https://www.hawking.org.uk/in-words/lectures/does-god-play-dice.

[113] Niels Bohr's retort to Einstein's comment on Max Born's analysis concerning the probability nature of quantum mechanics—"God does not play dice." This exchange took place during discussions on quantum theory at the Fifth Solvay International Conference in Brussels, Oct. 24–29, 1927.

[114] Bruce Rosenblum and Fred Kuttner, *Quantum Enigma: Physics Encounters Consciousness, Oxford University Press,* 2nd edition, 2011.

[115] Ibid.

[116] Op. cit., Francis Thompson, "The Mistress of Vision" poem.

[117] Stargate Project (US Army Unit), Wikipedia.

[118] J. D. Jost et al., Entangled Mechanical Oscillators, *Nature* [459, 683-685 (2009)].

[119] Jane Roberts, *The Unknown Reality,* TarcherPerigee Publications, July 8, 2011.

[120] https://plato.stanford.edu/entries/bell-theorem/.

[121] https://mickey-cheatham.blogspot.com/2013/04/.

[122] Freire Junior, O. Alain Aspect's Experiments on Bell's Theorem: A Turning Point in the History of the Research on the Foundations of Quantum Mechanics, *Eur. Phys. J. D* 76, 248 (2022). https://doi.org/10.1140/epjd/s10053-022-00542-z.

[123] Abide by Reason, Bell's Theorem Clearly Explained.

[124] Davide Castelvecchi and Elizabeth Gibney, "Spooky" Quantum-Entanglement Experiments Win Physics Nobel, https://www.nature.com/articles/d41586-022-03088-7.

[125] Gabriel Popkin, China's Quantum Satellite Achieves 'Spooky Action' at Record Distance: Result paves way for hack-proof quantum communications, *Science*, June 15, 2017.

[126] In theoretical physics, a No-Go Theorem is a theorem stating that a particular situation is not physically possible.

[127] Amanda Gefter, "Metaphysical Experiments" Probe Our Hidden Assumptions about Reality, *Quanta Magazine*, July 30, 2024.

[128] John Wheeler, *Today in Science History.*

[129] I used this concept in my sci-fi novel, *Cosmic Contact: The Next Earth—a New Beginning,* Waterside Productions, Cardiff-by-the-Sea, CA, March 5, 2023.

[130] Anton Zellinger, Quantum Teleportation, *McGraw-Hill Encyclopedia of Science & Technology*, vol. 14, pp. 705–06, 10th Edition, McGraw-Hill, New York, 2007.

[131] Op. cit., Helen Schucman.

[132] Feynman is referring to a process called *quantum tunneling*, which is important in the semiconductor industry. Quantum tunneling occurs when a subatomic particle or photon is able to penetrate through a very thin potential energy barrier that is higher in energy than the particle's kinetic energy. Tunneling commercial applications include the tunnel diode, quantum computing, flash memory, and the scanning tunneling microscope.

[133] Tara—Yoga Tantra Spirituality, the Connection Between Spirituality and Quantum Physics, Aug. 30, 2022.

[134] The Higgs field is a field of energy (consciousness can create both matter and energy) that Transcendent Physics posits exists in every region of the universe. The field is accompanied by a fundamental particle known as the Higgs boson, which is used by the field to continuously interact with other particles, such as electrons, and therefore with all other matter—hence, universal gravity.

[135] Op. cit., Transubstantiation.

[136] Op. cit., Nucleosynthesis.

[137] Op. cit., Our Best Understanding of the Universe May Be Wrong.

[138] An interesting historical point: During the medieval ages, charlatans posing as alchemists often milked unwitting nobles of funds with the promise of developing a means to transform base metals into gold or the creation of elixirs that prolong life. However, true alchemists were secretive, solitary, and spiritually inclined. They sought to uncover the deep secrets of nature and the universe to improve the overall quality of life. This encompassed two primary areas: metallurgy and medicinals. To achieve successful outcomes, prior to experimentation, these spiritually adept ancient philosophers and scientists frequently achieved high levels of consciousness via deep meditation. Much of their research was cast in secret symbols to avoid retribution by religious *Inquisitional* forces for their venturing into spiritualism and what might be called *metaphysics.*

[139] Meditation: For beginners, I suggest accessing *Deepak Chopra's 21-Day Meditation* called *Creating Abundance*, at https://www.deepakchopra.com/. To work at achieving deeper meditation and higher levels of consciousness, see James A. Cusumano, *Life Is Beautiful: 12 Universal Rules*, Chapter 8, *The Art and Science of Meditation*, pp. 87–99, Waterside Productions, Cardiff-by-the-Sea, CA, 2015.

[140] There are numerous means to elevate consciousness—some ancient, others modern extensions of these means to altered states. The most important point in personal pursuit of an effective practice is that the recipient be comfortable with, and passionate about, the practice of choice.

[141] A striking comment from a man of significant physical strength and accomplishment, who, in a serious accident, lost complete use of his body and spent the rest of his life interacting and communicating for the benefit of the *whole* solely through awareness and consciousness.

[142] Rupert Sheldrake, *Morphic Resonance: The Nature of Formative Causation,* 4th Edition, September 9, 2009, Park Street Press, South Parus, ME.

[143] Rupert Sheldrake estate.

[144] Wikipedia, *Von Neumann–Wigner interpretation* of the wave equation collapse upon observation.

[145] This is not referring to a multiverse model.

[146] The lowest level of energy left in the universe, even at absolute zero, available in near-infinite abundance.

[147] Dave Darling, Zero Point Energy, YouTube, Oct. 9, 2023.

[148] Zero Point Energy, Powering the Next Era of Humanity, YouTube, Jan. 1, 2025.

[149] *Scientific American*, What Is Zero-Point Energy?, Aug. 18, 1997.

[150] Psychic spies: Inside the government's secret ESP program, Aug. 19, 2018 / 9:00 AM EDT / CBS News.

[151] Russell Targ, Harold E. Putoff, *Mind-Reach: Scientists Look at Psychic Abilities (Studies of Consciousness)*, Hampton Roads Publishing, Feb. 1, 2005.

[152] Andy Fell, University of California-Davis, Psychic Spying Research Produces Credible Evidence, Nov. 28, 1995.

[153] Jasmine Soumana, I Listened to 100 Near-Death Experiences—Earth Is Not What We Think It Is: 8 Lessons to Take Seriously While You're Still Alive, *Medium*, April 16, 2025.

[154] T. Lenton, *Gaia Hypothesis, Encyclopedia of Atmospheric Science*, pp. 815–20, Academic Press, New York, 2003.

[155] Op. cit., Meditation, *Deepak Chopra's 21-Day Meditation* called *Creating Abundance*, Meditation, No. 3.

[156] Isha Sadhguru Foundation, Nov. 2, 2021.

[157] Loyola Press.

About the Author

James A. Cusumano (Jim) is chairman and owner of Chateau Mcely, chosen in 2007 by the European Union as the only "green" five-star castle hotel in Central Europe; and in 2008 by the World Travel Awards as *"The World's Leading Green Hotel."* In 2025, the *Michelin Guide* recognized Chateau Mcely among a select group of hotels throughout the world. Their recently launched *Key Program* showcases hotels in the same manner that their *Star Program* has awarded restaurants of distinction over the last 125 years.

Chateau Mcely offers programs that promote the principles of Inspired and Conscious Leadership, finding one's Life Purpose, and Long-Term Longevity and Personal Fulfillment.

Jim began his career during the 1950s in the field of entertainment as a recording artist. Years later, after earning a PhD in experimental quantum physics at Rutgers University, studying business at Stanford University, and being designated a Foreign Fellow of Churchill College at Cambridge University in the UK, he joined Exxon as a research scientist and later became the research director for Exxon's Catalytic Science & Technology Group.

Jim subsequently cofounded two public companies (NASDAQ Stock Exchange) in Silicon Valley: Catalytica Energy Systems, Inc.—devoted to clean power generation; and Catalytica Pharmaceuticals, Inc., which manufactured drugs via environmentally benign, low-cost, catalytic technologies. While he was chairman and CEO, Catalytica Pharmaceuticals grew in less than five years from several employees to more than 2,000, and became a $1 billion enterprise before being sold.

Subsequent to his work in Silicon Valley and before buying and renovating Chateau Mcely with his wife, Inez, Jim returned to the entertainment industry and founded Chateau Wally Films, which produced the feature film *What Matters Most* (2001), distributed in more than 60 countries.

Jim's research decades ago in quantum physics created a lifelong passion to understand the connection between quantum mechanics and consciousness. Over many years, this eventually led to his efforts in Transcendent Physics.

Jim lives in Prague with his wife, Inez, and their daughter, Julia.

www.JimTheAlchymist.com